Machine L
and Analytics in Healthcare Systems

Green Engineering and Technology: Concepts and Applications

Series Editors
Brujo Kishore Mishra

GIET University, India, and Raghvendra Kumar, LNCT College, India

Environment is an important issue these days for the whole world. Different strategies and technologies are used to save and protect the environment. Technology is the application of knowledge to practical requirements. Green technologies encompass various aspects of technology, which help us to reduce the human impact on the environment and to create ways of sustainable development. This book series will describe innovations in green technology in different ways, aspects, and methods. This technology helps people to understand the use of different resources to fulfill needs and demands. Some points will be discussed in terms of the combination of involuntary approaches and government incentives, while a comprehensive regulatory framework will encourage the spread of green technology, with least developed countries and developing states requiring unique support measures in order to promote green technologies.

Cognitive Computing Using Green Technologies
Modeling Techniques and Applications
Edited by Asis Kumar Tripathy, Chiranji Lal Chowdhary, Mahasweta Sarkar, and Sanjaya Kumar Panda

Handbook of Green Engineering Technologies for Sustainable Smart Cities
Edited by Saravanan Krishnan and G. Sakthinathan

Green Engineering and Technology
Innovations, Design, and Architectural Implementation
Edited by Om Prakash Jena, Alok Ranjan Tripathy, and Zdzislaw Polkowski

Machine Learning and Analytics in Healthcare Systems
Principles and Applications
Edited by Himani Bansal, Balamurugan Balusamy, T. Poongodi, and Firoz Khan KP

Convergence of Blockchain Technology and E-Business
Concepts, Applications, and Case Studies
Edited by D. Sumathi, T. Poongodi, Bansal Himani, Balamurugan Balusamy, and Firoz Khan K P

For more information about this series, please visit: https://www.routledge.com/Green-Engineering-and-Technology-Concepts-and-Applications/book-series/CRCGETCA

Machine Learning and Analytics in Healthcare Systems

Principles and Applications

Edited by
Himani Bansal, Balamurugan Balusamy,
T. Poongodi, and Firoz Khan KP

CRC Press
Taylor & Francis Group
Boca Raton London New York

CRC Press is an imprint of the
Taylor & Francis Group, an **informa** business

First edition published 2021 by
CRC Press
6000 Broken Sound Parkway NW, Suite 300, Boca Raton, FL 33487-2742

and by
CRC Press
2 Park Square, Milton Park, Abingdon, Oxon, OX14 4RN

CRC Press is an imprint of Taylor & Francis Group, LLC

Library of Congress Cataloging-in-Publication Data

Names: Himani, Bansal, editor. | Balusamy, Balamurugan, editor. | Poongodi, T., editor. | Khan KP, Firoz, editor.
Title: Machine learning and analytics in healthcare systems : principles and applications / edited by Bansal Himani, Balamurugan Balusamy, T. Poongodi, and Firoz Khan KP.
Description: First editlion. | Boca Raton : CRC Press, 2021. | Series: Green engineering and technology: concepts and applications | Includes bibliographical references and index.
Identifiers: LCCN 2021000895 (print) | LCCN 2021000896 (ebook) | ISBN 9780367487935 (hardback) | ISBN 9781003185246 (ebook)
Subjects: LCSH: Machine learning. | Medical informatics.
Classification: LCC Q325.5 .M32158 2021 (print) | LCC Q325.5 (ebook) | DDC 006.3/1--dc23
LC record available at https://lccn.loc.gov/2021000895
LC ebook record available at https://lccn.loc.gov/2021000896

ISBN: 978-0-367-48793-5 (hbk)
ISBN: 978-0-367-56486-5 (pbk)
ISBN: 978-1-003-18524-6 (ebk)

Typeset in Times
by Deanta Global Publishing Services, Chennai, India

Contents

Preface

"It is not the strongest of the species that survives, nor the most intelligent, but the one most responsive to change."

- Charles Darwin

By augmenting the human performance, machine learning and analytics have the potential ability to restore "care" to health care. Health care is one of the crucial subject areas. Detection of pertinent information is a critical task. Machine learning and health analytics has placed the patient at the center of the healthcare ecosystem. Patients, the people seeking increased wellbeing and health, have been provided with a plethora of technology-oriented features, with machine learning and analytics at its core. The health-related data are also handled at a lightning speed. The onus is to find nuggets of information from this vast ocean of data, which is unlikely to be achieved without these cutting-edge technologies. Furthermore, machine learning and analytics are being integrated into all the sectors generating huge numbers of data points every day. This edited book covers all aspects of the machine learning techniques and applications, with a complete 360-degree view. This book seeks to close the gap between engineering and medicine, combining health sciences with the design and problem-solving skills of engineering to advance healthcare treatments, including diagnosis, monitoring, and therapy.

Health plays such a vital role in our lives, that healthcare management becomes an essential part of society. Traditionally, healthcare was considered to be the domain of only medical practitioners. But, as the 'digital age' has approached, the medical field have become digitized too. Machine learning and analytics have given wings to the healthcare industry. Healthcare solutions, when powered by machine learning and analytics, increase as well as deepen. Several case studies, combining machine learning with health care, are also incorporated in this book. The wide variety of topics it presents offers readers multiple perspectives on a range of disciplines.

Chapter 1 by Suganthi et al. details the principles, challenges, and applications relating to data analytics in healthcare. They touch on the viability of handling enormous volumes of heterogeneous data and of evidence-based decisions being taken to improve the healthcare services. The tools and techniques used are also detailed.

Chapter 2 by Narayanan et al. entails a systematic analysis of machine learning in healthcare systems. The role of artificial intelligence (AI) is unprecedented in this field. They discuss different AI classifiers for different dynamic areas.

Chapter 3 by Babu discusses the groundwork of machine learning-based data classification techniques used in health care. Various supervised and unsupervised machine learning algorithms, used in health care for a range of applications, are described.

Chapter 4 by Sreenarayanan et al. takes the insights of this book farther by exploring deep learning techniques for computer-aided medical diagnosis. They ponder upon the changes that have led to the penetration of deep learning techniques into health care, the major deep learning models used, and their applications.

Chapter 5 by Sambath Kumar and Rajendran stresses that automatic classification plays a major role in the medical field. They discuss various case studies of health care (COVID-19, SARS, and tumors) related to machine learning classifiers.

Chapter 6 by Jagadeesh and Rajendran discusses various machine learning approaches for analysis in healthcare informatics. They also use deep learning and natural language processing, along with other machine learning approaches, for their analysis.

Chapter 7 by Gupta and Kumar Prasad discusses the application of machine learning techniques to the healthcare sector for the prediction of epidemic disease outbreaks. They also highlight how machine learning techniques can be used to study the current outbreak of COVID-19.

Chapter 8 by Kokilavani and Beena covers different case studies, such as the health monitoring system, disease prediction, precision medicine, and clinical support system for better prediction of diseases, using machine learning methods to aid healthcare treatment techniques.

Chapter 9 by Malviya and Sharma details the applications of computational methods and modeling in drug delivery. They show how this can be used in various phases of drug discovery and development, thus reducing the time and cost involved.

Chapter 10 by Annapurani et al. discusses the role of a business intelligence tool in healthcare data analytics, which can be used for prediction of infectious diseases and to develop a recommendation system. It also helps in identifying and handling of outlier data.

Chapter 11 by Seraphim and Poovammal underlines the use of machine learning-based data classification techniques on different data types (sensor data, clinical data, and omics data). They focus on how classification techniques and predictive tools can be used in the exact prediction of results.

Chapter 12 by Ghantasala et al. describes the disease health monitoring of COVID-19 with a clinical support system and its objectives.

Himani Bansal
Balamurugan Balusamy
T. Poongodi
Firoz Khan KP

Editors

Himani Bansal has over 14 years of experience in academia and the IT industry. She is currently working as an Assistant Professor in Jaypee Institute of Information Technology, Noida, India, and possesses many reputed certifications, such as UGC National Eligibility Test (NET), IBM Certified Academic Associate DB2 9 Database and Application Fundamentals, Google Analytics Platform Principles by Google Analytics Academy, E-Commerce Analytics by Google Analytics Academy, and RSA (Rational Seed Academy) and SAP-ERP Professional. Her research interests include machine learning and data analytics, cloud computing, business analytics, data mining, and information retrieval. Dr Bansal has filed four patents and has around 40 publications, including edited books, an authored book, peer-reviewed papers in international journals, and abstracts in proceedings of conferences of high repute. She has served as section editor, guest editor, convener and session chair for various highly regarded journals and conferences, such as SCPE, NGCT, IndiaCom, CSI Digital Life, IJAIP, JGIM, ICACCI, ICCCA, etc., and has reviewed many research papers for prestigious journals. Dr Bansal serves as a Life Member of various professional societies such as CSI, ISTE, CSTA, and IAENG, and is an active member of IEEE and ACM. Recently, IEEE has conferred her with Senior Membership.

Balamurugan Balusamy has reached the position of Associate Professor in his 14-year career to date at VIT University, Vellore, India. He has completed his Bachelors, Masters and Ph.D. degrees from premier institutions. His passion is teaching and he adopts different design thought principles while delivering his lectures. Dr Balusamy has written around 30 books on various technologies and visited more than 15 countries during his work. He has presented his research at many important conferences and has published over 150 peer-reviewed papers, book chapters, and conference proceeding abstracts. Dr Balusamy serves on the advisory committees for several start-ups and forums, and does consultancy work for industry on Industrial IoT. He has given over 175 talks at various events, conferences, and symposia. He is currently working as Professor in Galgotias University in India, combining teaching with research on block chain and IoT.

T. Poongodi is an Associate Professor in the School of Computing Science and Engineering, Galgotias University, Greater Noida, India. She completed her Ph.D in Information Technology (Information and Communication Engineering) from Anna University, Tamil Nadu, India. The main focuses of her research are Big Data, Internet of Things, ad-hoc networks, network security, and cloud computing. Dr Poongodi is a pioneering researcher in the areas of Big Data, wireless networks, and Internet of Things, and has published more than 25 peer-reviewed papers in various international journals. She has presented papers at national and international conferences, written chapters in books published by CRC Press, IGI Global, Springer, and Elsevier, and has edited books published by CRC, IET, Wiley, Springer, and Apple Academic Press.

Firoz Khan KP was born in Kerala, India, in 1974. He received his BSc degree in Electronics from the Bharatiyaar University, Coimbatore, India, in 1991 and his Masters degree in Information Technology from the University of Southern Queensland, Australia, and a second Master's degree, in Information Network and Computer Security (with Honors), from the New York Institute of Technology, Abu Dhabi, UAE, in 2006 and 2016, respectively. He is currently working towards his PhD in Computer Science from the British University in Dubai, Dubai, UAE. In 2001, he joined the Higher Colleges of Technology, Dubai, UAE in the Computer Information Science department as a Teaching Technician and progressed to become a Faculty member in 2005. He is currently a lecturer, with security and networking being his primary areas of teaching. His current research fields include computer security, machine learning, deep learning, and computer networking.

Contributors

Annapurani K.
Department of Computer Science and Engineering
School of Computing
SRM Institute of Science and Technology
Kattankulathur, India

Arjun K. P.
School of Computing Science and Engineering
Galgotias University
Greater Noida, India

M. Arvindhan
School of Computing Science and Engineering
Galgotias University
Greater Noida, India

L. Godlin Atlas
School of Computing Science and Engineering
Galgotias University
Greater Noida, India

Bindu Babu
Department of Electronics and Communication Engineering
Easwari Engineering College
Ramapuram
Chennai, India

T. Lucia Agnes Beena
Department of Information Technology
St. Josephs College
Tiruchirappalli, India

Deepa C. M.
Computer Science and Engineering
Indian School Al Maabela
Muscat, Sultanate of Oman

G. S. Pradeep Ghantasala
Computer Science and Engineering
Punjab Campus
Punjab, India

Vaishali Gupta
School of Computing Science and Engineering
Galgotias University
Greater Noida, India

B. Ida Seraphim
Department of Computer Science and Engineering
School of Computing
SRM Institute of Science and Technology
Kattankulathur, India

Jagadeesh K.
SNS College of Engineering
Anna University
Tamil Nadu, India

K. Sambath Kumar
Vel Tech Rangarajan Dr. Sagunthala R&D
Institute of Science and Technology
Chennai, India

T. Kokilavani
Department of Computer Science
St. Josephs College
Tiruchirappalli, India

Rishabha Malviya
School of Medical and Allied Sciences
Galgotias University
Greater Noida, India

Neethu Narayanan
School of Department of Vocational Studies (B.Voc Software Development)
St. Mary's College
Calicut University
Thrissur, India

Sanjeev Kumar Prasad
School of Computing Science and Engineering
Galgotias University
Greater Noida, India

T. Poongodi
School of Computing Science and Engineering
Galgotias University
Greater Noida, India

E. Poovammal
Department of Computer Science and Engineering
School of Computing
SRM Institute of Science and Technology
Kattankulathur, India

A. Rajendran
Karpagam College of Engineering
Myleripalayam Village
Othakkal Mandapam, India

Anu Radha Reddy
Computer Science and Engineering,
Malla Reddy Institute of Technology and Science-(NH 45)
Telungana, India

Caroline Ruvinga
Department of Computer Science
Midlands State University
Gweru, Zimbabwe

Akanksha Sharma
School of Medical and Allied Sciences
Galgotias University
Greater Noida, India

Sreenarayanan N. M.
School of Computing Science and Engineering
Galgotias University
Greater Noida, India

S. Sudha
Department of Electronics and Communication Engineering
Easwari Engineering College
Chennai, India

S. Suganthi
PG and Research Department of Computer Science
Cauvery College for Women
Tiruchirapalli, India

Varsha Sisaudia
School of Computing Science and Engineering
Galgotias University
Greater Noida, India

Ibrahim Venkat
School of Computing and Informatics
Universiti Teknologi Brunei
Gadong, Brunei Darussalam

1 Data Analytics in Healthcare Systems – Principles, Challenges, and Applications

S. Suganthi, Vaishali Gupta, Varsha Sisaudia, and T. Poongodi

CONTENTS

1.1 INTRODUCTION

The healthcare industry is multidimensional, with multiple data sources involving healthcare systems, health insurers, clinical researchers, social media, and government [1], generating different types and massive amounts of data. It is impossible to handle this big data with traditional software and hardware and the existing storage methods and tools. Data analytics is the process of the analysis of data to identify trends and patterns to gain valuable insights. The data generated in the health industry are characterized by the four Vs of big data, namely volume, velocity, variety, and veracity, which play crucial roles in health data analytics. Also, evidence-based decision making has gained importance, which involves the sharing of data among various data repositories. According to Deloitte Global Healthcare Outlook, it is expected that global healthcare expenditure will continue to increase at an annual rate of 5.4% between 2017 and 2022. This is due to the increased importance of personalized medicine, the use of advanced technologies, the demand for new payment models, improvement and expansion of care delivery sites, and competition. Various research attempts, based on big data, have provided strong evidence that the efficiency of healthcare applications is dependent upon the basic architecture, techniques, and tools used. Statistical data and reports can be generated with the use of patient records, aiding in knowledge discovery, and thereby influencing value-added services to the patients, improving healthcare quality, the making of timely decisions, and minimizing the costs incurred. Hence, there is a need to incorporate and integrate big data analytics into existing healthcare systems. Despite healthcare analytics having massive potential for value-added change, there are many technological, social, organizational, economic, and policy barriers associated with its application [2].

1.1.1 Data Analytics in Healthcare

Health industries employing data analytics can use big data for the early detection of diseases and their treatment, clinical operations, genomic analysis, patient profile analytics, and prevention of fraud. Data processing involves analytics being applied to the transformed data to obtain meaningful insights from the healthcare data for evidence-based decision making. There exist three levels of analytics with increasing complexity and value, which are given below.

- **Descriptive Analytics** describes the current events or summarizes the past events by generating reports with the help of statistical tools such as tables and graphs. Thus, it helps medical practitioners to study and understand the patient's behavioral patterns in the past with the help of the patient's operational data, which can be used to solve problems in current situations.
- **Predictive Analytics** enables the user to predict the future with the use of descriptive data, using empirical methods, such as machine learning, modeling, and data mining, to analyze the data.
- **Prescriptive Analytics** enables the user to choose the best solution from several possible alternatives to the issues in question. The real-time data are analyzed with the historical data by the use of artificial intelligence, data mining, and machine learning. The practitioners can determine the possible effects of their decisions by using simulation and optimization techniques and can prepare themselves to handle any event in case of failure or success.

1.1.2 Characteristics of Big Data

The main characteristics of big data are the four Vs, which are as follows.

- **Volume:** The healthcare industry generates an enormous amount of data coming from various sources, such as EHRs (Electronic Health Records), LIMS (Laboratory Information Management System), diagnostic or monitoring instruments, supply chains, insurance claims/billing, pharmacy, real-time locating systems [3], and social media. The data that are collected are used in continuous learning by employing various technologies and processes to derive insights from the information, which will help improve the quality of healthcare. In addition, the reduction in storage costs and the development of advanced architectures have led to large volumes of data being stored, processed, and managed, using the existing traditional systems adopted by the healthcare industry.
- **Velocity:** The speed with which medical data are generated is high and requires specific processing requirements. The healthcare industry has not adapted to technological advances, and hence various processing methods have been adopted to cope with the speed with which the data are produced. Batch processing, stream processing, near-real-time processing and real-time processing methods are used to handle data in the healthcare industry [4].
- **Variety:** Data in healthcare come from several sources and are in various forms. About 80% of the total healthcare data is unstructured (e.g., images, signals, text), which does not fit into any predefined data format, data type, or structure. The remaining 20% of the data is structured (e.g., temperature, blood pressure, patient demographics, etc.), possessing a predefined data format. The structured data are easy to handle, store, process, and manipulate. Hence, the unstructured data must be converted to a structured form by efficient automatic transformation methods. When combined with structured data, the unstructured data provide valuable information, which can be harnessed to improve value-added services in healthcare by adopting more-efficient methodologies.
- **Veracity:** This refers to the accuracy of the data collected, which is directly proportional to the value of the insights which can be obtained from them. The results derived from data analytics are error free if the data obtained are trustworthy, accountable, authenticate, and available.

1.2 ARCHITECTURAL FRAMEWORK

Hadoop/MapReduce is an open-source platform for big data analytics used in healthcare, which performs parallel processing in a distributed environment, involving multiple nodes in the network. The use of Hadoop and MapReduce technologies has been found to be fruitful in many healthcare applications, by improving the performance of, for example, image processing, neural signal processing, protein

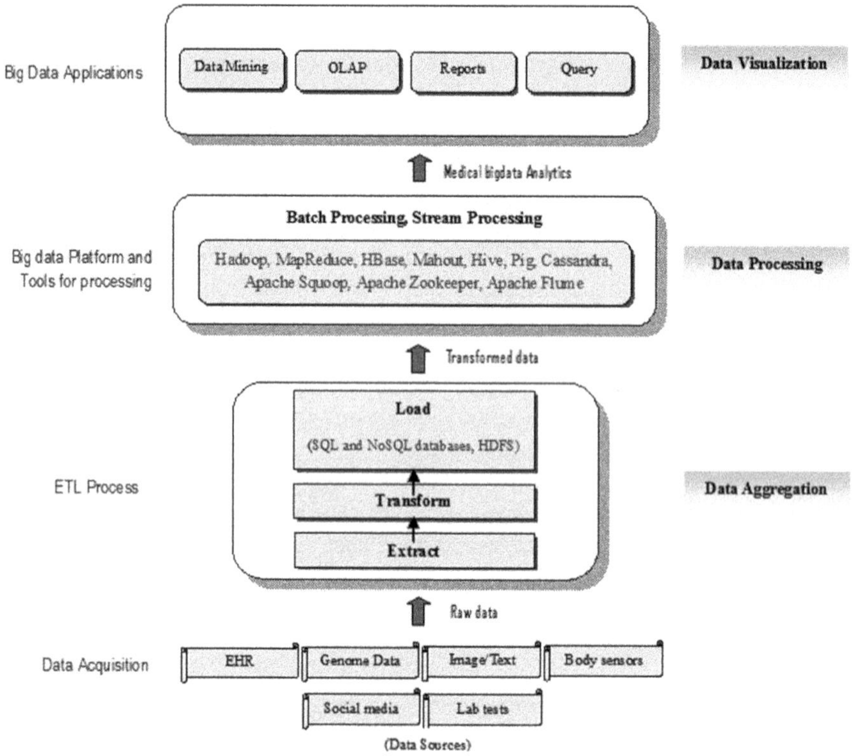

FIGURE 1.1 Conceptual framework of big data architecture in healthcare.

structure alignments, signal detection algorithms, and lung texture classification [5]. The architectural framework of big data in healthcare is composed of three major components, namely data aggregation, data processing, and data visualization [6]. Figure 1.1 illustrates the conceptual framework of big data architecture in healthcare.

1.2.1 Data Aggregation

Data aggregation in healthcare involves the process of collecting and integrating raw data from various modalities and multiple systems and converting them into a single standard format suitable for analysis, processing, and storage in a data warehouse. The functionalities involved in the process include data extraction, data transformation, and data loading.

- **Data Extraction**
  Healthcare data occupy large volumes and come from heterogeneous sources. The primary sources of medical data include medical records, health surveys, claims data, disease registries, vital records, surveillance data, peer-reviewed literature, clinical trial data, and administrative

records. The data from these sources can be structured or unstructured. The structured data, which contain numeric, categorical, and nominal data types, have a predefined format and are easy to handle. Unstructured medical data, such as image data from imaging devices, text data (e.g., doctor notes, shorthand notations etc.), and signal data (e.g., biosignals from wearable devices), do not have a predefined format. They must be converted into a standard format for further processing. The medical data can be in any format, which includes EHRs, images, biomedical data, genomic data, sensor data, and clinical text data. These data are in the form of text/ASCII, XML, JSON, or images in DICOM formats. Usually, medical image data from local workstations are stored in PACS (Picture Archiving and Communication Systems) and are transferred and communicated to other workstations following DICOM standards.

- **Data Transformation**
  The acquired raw data are transformed in order to apply particular business logic by passing through the phases of *data filtering*, which is the process of removing unwanted data ("data cleaning"), which includes the processes of normalization, noise reduction, and management of missing values, and *data manipulation*. The data converted in this way are in a standard format that is consistent and suitable for further analysis.
- **Data Loading**
  Customized relational databases are usually designed for each particular healthcare system, with its own defined data models and schemes for the storage of medical data. The transformed medical data, which are suitable for further analytics, are loaded into the target database or a data warehouse, such as HDFS (Hadoop Distributed File System), SQL relational databases, or NoSQL databases, or combinations of these.

1.2.2 Data Processing

The data processing used in healthcare includes *batch processing* and *stream processing* methods [7]. Batch processing is the method of analyzing data in batches, which are collected and stored over a period and in which response time is not considered. On the other hand, stream processing is the method of analyzing huge volumes of data, to which a real-time response is required. Some applications in healthcare require real-time processing of data and they are characterized by noisy data with missing or redundant values, continuous changes in data, and the need for a rapid response. Stream processing overcomes these difficulties with simple and rapid information extraction by using data-mining methods, such as clustering, classification, and frequent pattern mining [7]. Apache Hadoop MapReduce is a popular framework used for batch processing, whereas Storm and S4 are frameworks used for stream processing, with Apache Spark and Apache Flink being frameworks used for both batch and stream processing.

The Hadoop platform is most widely used for batch processing in which parallel processing of huge volumes of data are carried out in a distributed manner. It is a framework in which the process of big data analytics is conducted through a collection of various tools, methodologies, and libraries. It consists of two main components, the HDFS and Hadoop MapReduce.

- **Hadoop Distributed File System (HDFS)**
  The HDFS is a distributed file system used for storing and retrieving extremely large volumes of data at great speed. The data are split into several blocks with uniform block sizes of 64 MB or 128 MB, which are distributed across many server nodes, enhancing parallel processing. The HDFS consists of two nodes, namely a *NameNode* and a multiple number of *DataNodes*. The DataNodes contain the application data and the NameNodes contain metadata related to the storage and retrieval of data from the DataNodes. The data are replicated and the copies are distributed to many nodes, making the system fault tolerant in the case of any node failure.
- **MapReduce**
  MapReduce is a framework and programming model for distributed processing of huge datasets, involving multiple nodes. The processing is broken down across several individual nodes and carried out in parallel because the data are vast. The process consists of two phases, the map phase and the reduce phase, conducted with the mapper function and the reducer function, respectively.
  1. **Map Phase:** the data stored in the HDFS are split into smaller, fixed-sized elements and passed to a mapper function. The mapper function produces structured output data tuples of each component in the form of key/value pairs, which are written into an intermediate file.
  2. **Reduce Phase:** the output from the map phase is passed to the reduce phase, before which it is shuffled to consolidate and combine the relevant data tuples into smaller ones. The reducer aggregates the output from shuffling by merging the same keys to form a smaller set of tuples, which is written to a single output file.

Data storage can be carried out with tools other than HDFS, such as HBase, Hive, Cassandra, Pig, Apache Flume, Apache Squoop, and other relational databases. Whereas Apache Oozie is used in the case of large numbers of interconnected systems, Apache Zookeeper is used to maintain application reliability, and Mahout is used for machine- learning purposes [7].

1.2.3 Data Visualization

Visualization is the graphical representation of data, which helps the practitioner gain more insights from the data. The analytical tool cleanses and evaluates the data with the help of data-mining algorithms, evaluation, and software tools before the data are visualized. The main applications of data analytics to healthcare are queries, reports, online analytical processing (OLAP), and data mining. They are used for displaying predictive reports, proactive messages, visualization of patient health records, real-time alerts, and dashboards for monitoring the daily health status of patients.

1.3 DATA ANALYTICS TOOLS IN HEALTHCARE

One of the critical challenges in healthcare systems is utilizing the massive amount of data generated daily in an efficient and cost-effective manner. Hence, healthcare systems require effective and efficient tools to ensure the appropriate use of the data. Figure 1.2 depicts the data analytics tools used in healthcare.

1.3.1 Data Integration Tools

Data integration in healthcare applications refers to the act of combining health data from a myriad of sources into a unified set of accumulated data that provide actionable business intelligence. Integration of multiple medical databases can be useful in identifying different methods of disease prevention, providing more sophisticated and personalized care, and reducing costs by avoiding overuse of resources.

As health data come from multiple, disparate sources like medical devices, wearable devices, etc., healthcare professionals face a major challenge in dealing with these unstructured data. Collecting and consolidating health data from various sources is extremely beneficial but still poses a major challenge to the health sector. The process of integrating huge volumes of medical data is complicated and presents significant challenges. Some of these challenges include:

- Lack of standard data formats
- Data privacy and confidentiality regulations
- Data format inconsistency among various healthcare applications
- Need for greater integration processing power
- Low end-user adoption

However, data integration tools can be used to integrate health data from multiple sources to generate meaningful insights from the data. Data integration tools include software and platforms that can aggregate data from disparate sources. The

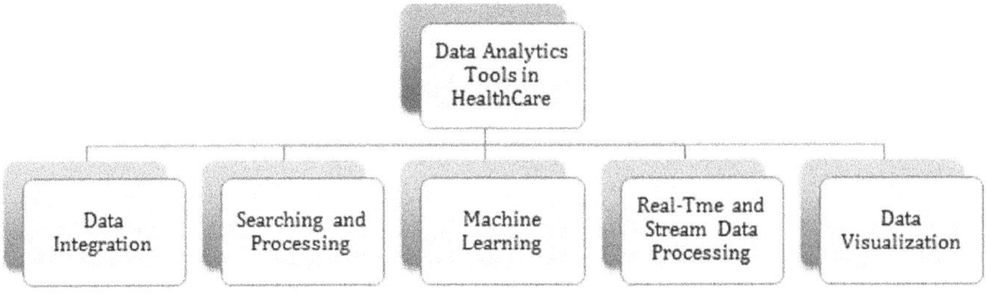

FIGURE 1.2 Data analytics tools in healthcare.

following are some data integration solutions that can be considered in healthcare organizations to make better and more efficient use of healthcare data.

- **Attunity** is an integration tool that can aggregate disparate data and files across all major databases, including cloud platforms, data warehouses, and Hadoop. It also supports the Health Level 7 (HL7) messaging standard, which is a healthcare standard. It can integrate and connect with web applications in real time.
- **Informatica** is an advanced, multi-cloud and hybrid data integration tool that can integrate data from multiple, disparate datasets, such as data warehouses, Hadoop, enterprise applications, message applications, and midrange systems. It also provides data management tools for companies in the healthcare field to facilitate patient services with improved outcomes and reduced costs. Informatica's cloud integration allows administrators to integrate data with in-house applications, claims processing, etc., in health organization environments.
- **Information Builder** is a data integration and business intelligence tool that can measure and aggregate very large healthcare data collected throughout the patient lifecycle. These tools ensure the availability of data in real time across the healthcare environment and enhance the quality of health services.
- **Jitterbit** [8] is a single, secure, cloud-based data integration platform that aggregates structured and unstructured health data or clinical data retrieved from sources such as EHR. It enables more-efficient operations and provides complete access to health data in a format that can be used with other systems.
- **Magic** is a data integration tool that connects disparate systems for healthcare organizations. It ensures the best possible care of patients by keeping all health-related records up to date and available to all healthcare providers. Magic's integration platform combines diverse systems into a single interface *via* the graphical user interface.

1.3.2 Searching and Processing Tools

Since healthcare organizations deal each day with a large volume of unstructured data, there is a great need for data indexing, searching, and processing tools to optimize the efficient use of clinical data . These tools are employed for effective management of data that is stored in HDFS, which allows multiple files to be stored and retrieved at the same time in a big data environment. The following are some of the searching and processing tools used in healthcare.

- **Lucene** [8] is a scalable tool for indexing large blocks of unstructured text that provides advanced, full-text search capabilities. It can integrate easily with Hadoop to facilitate distributed text management.

- **Google Dremel** [9] is a distributed system that uses multi-level execution trees for interactive query processing of large datasets.
- **Cloudera Impala** is an MPP (Massively Parallel Processing) SQL query tool that supports the accessing of a massive volume of data stored in Hadoop clusters. It is a scalable and flexible parallel database technology that enables users to directly query data stored in HDFS file formats without requiring data format transformation.
- **Apache Hive** is a database query tool which facilitates querying, writing, and managing large datasets stored in distributed storage using a HiveQL language, which is a SQL-like query language. It allows structure to be projected onto data already in storage.

1.3.3 Machine Learning Tools

In a healthcare environment, machine learning tools are used to convert comprehensive health data into actionable knowledge that supports effective decision making to perform informed clinical activities [8].

- **Apache Mahout** [10] is an open-source, powerful, scalable machine learning library that runs on top of Hadoop MapReduce. The Mahout library facilitates the execution of distributed or scalable machine learning algorithms. The Mahout algorithms mainly focus on classification, clustering, and collaborative filtering techniques.
- **Skytree** is a big data machine learning tool that offers more accurate and faster predictive analytics models that are easy to use. These models enhance the processing of massive datasets in a precise manner without down sampling.
- **Apache SAMOA** (Scalable Advanced Massive Online Analysis) is an open-source-distributed, streaming machine learning framework that enables users to create distributed, streaming machine learning algorithms and to execute them on multiple DSPEs (Distributed Stream Processing Engines) [11].
- **BigML** [8] is a scalable, open-source machine learning tool that provides a framework to perform sophisticated machine learning workflows such as classification, regression, cluster analysis, anomaly detection, and association discovery. It includes a cloud infrastructure that can readily be integrated with machine learning features to build cost-effective, scalable, flexible, and reliable applications.

1.3.4 Real-Time and Streaming Data Processing Tools

With the rapid growth of massive Internet of Things (IoT), including wearables and medical devices in smart healthcare systems, there is a need for real-time and streaming data processing tools to carry out real-time processing of health data.

- **Apache Storm** [12] is an open-source, real-time data processing platform to process streaming data in real time. It also provides a fault tolerant, real-time computational framework. The applications of Apache Storm include real-time analytics, log processing, ETL jobs, continuous computation, distributed real-time processing, and machine learning.
- **SQL Stream Blaze** [12] is a streaming analytics platform that supports a variety of available streaming sources in all formats and at all speeds. Applications of SQL streaming include high throughput, data discovery, data wrangling, real-time threat detection, and analytics.
- **Apache Flink** [12] is a distributed processing engine, and it provides a framework for stateful computations over bounded and unbounded data streams. It is capable of both batch and stream processing and offers efficient, fast, accurate, and fault-tolerant handling of massive streams of data.
- **Apache Kafka** is an open-source, distributed event-streaming platform that provides a framework for building high-performance real-time streaming data pipelines, streaming analytics, and applications. There are five core application programming interfaces (APIs) in Kafka, namely Producer API, Consumer API, Streams API, Connector API, and Admin API, to facilitate message passing, storage, and stream processing services.

1.3.5 Visual Data Analytical Tools

Data visualization is a pictorial or graphical representation of data that enables users in any organization to analyze and understand the trends or patterns of data. In healthcare, data visualization tools help in creating and monitoring the dynamic data of a patient, presenting clinical records, identifying patterns and trends, and carrying out time-series analysis to improve healthcare services and public health policy.

- **SAS Visual Analytics** [13] is a web-based analytical tool that allows multiple users to access a massive amount of real-time data simultaneously from a LASR analytical server. It allows parallel networking by transferring data from one machine to another machine to access secure data quickly.
- **Tableau** [13] is a business intelligence visualization tool that transforms raw and large datasets into a defined format to provide real-time structured data to support decision making. One of the uses of Tableau has been to quickly diagnose genetic diseases and to help health practitioners in providing rapid treatment to the patients.
- **QlikView** [13] is a visualization tool that transmits related data from different sources into electronic medical records. It provides in-memory analysis and reduces the risk of medical error by tracking safety metrics and lowers the cost of delivering services to the patients. It ensures that all regulatory compliance in the healthcare system is delivered and maintained in a timely manner.

1.4 DATA ANALYTICS TECHNIQUES IN HEALTHCARE

Healthcare big data refers to multidimensional health data amassed from various sources, including medical imaging (X-ray, magnetic resonance imaging (MRI), and computed tomography (CT) scan images), structured data EHRs, biomedical signals (ECG, EEG, etc.), handwritten prescriptions, and data from wearables and medical devices. Since health data are dynamic and complex, they are difficult to manage and analyze using traditional techniques and technologies. There is a great demand for effective data analytics techniques to study these diverse data and to facilitate decision making. Figure 1.3 depicts the various big data analytics techniques used in healthcare, some of which are described below [5].

- **Data Mining** is useful for discovering patterns and for extracting meaningful information from large databases. With the rapid growth in massive health data, data-mining techniques have helped to search for new and valuable information (knowledge) from large complex databases in

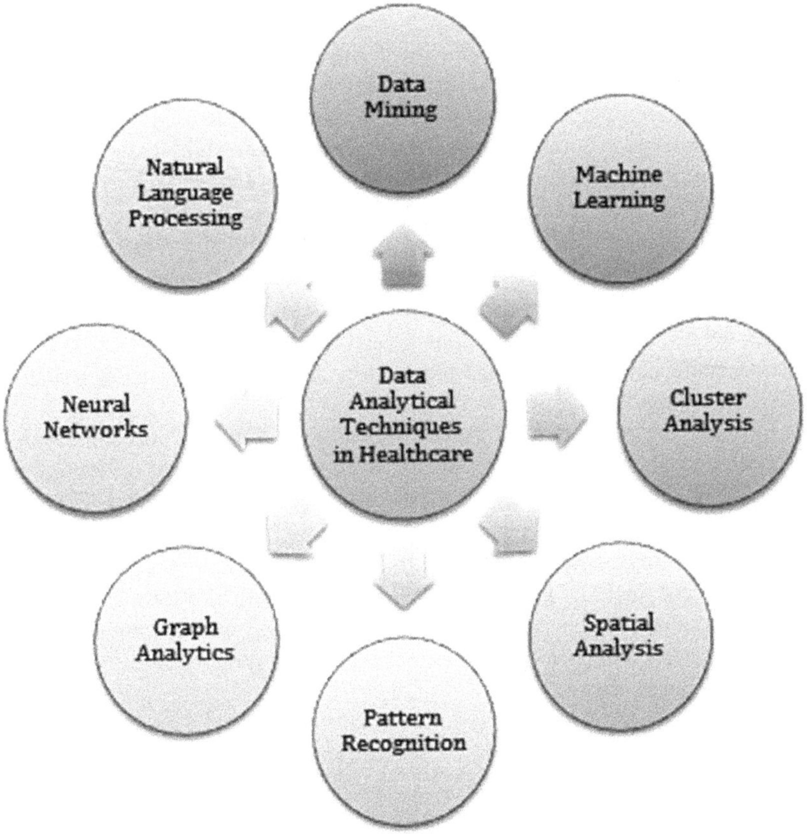

FIGURE 1.3 Data analytics techniques in healthcare.

healthcare systems, that facilitate the decision-making process. The main application areas of data mining include prediction and determination of various diseases, biosignal monitoring of patients, assistance in diagnosis and treatments, and exploratory data analysis in healthcare.

- **Machine Learning (ML)** in healthcare helps to analyze data and suggest outcomes. Applications of ML include prediction of diseases and diagnosis, drug discovery and manufacture, hospital performance assessment, smart health records, personalized patient care, etc.
- **Cluster Analysis** helps to discover hidden structures and clusters found in massive databases. In healthcare, cluster analysis helps to identify subgroups in the patient population defined by the patient's characteristics, disease severity, and treatment responses. Cluster analysis can be used, for example, to determine obesity clusters, to identify high-risk patient groups, and to identify population clusters with specific disease determinants in order to optimize treatment.
- **Graph Analytics** techniques analyze data in the form of a graph, where individual data entities are represented as nodes and the relationships that connect those entities are represented as edges. Graph analytics are used in healthcare to estimate healthcare fraud risk and hospital performance analysis based on quality measures [14].
- **Natural Language Processing (NLP)** is the technique used to make computers understand human speech and text. In healthcare, NLP has great potential to search, analyze, and interpret large volumes of patient-related datasets. Applications of NLP include the provision of efficient patient care, control of health service cost, extraction of meaningful health-related data from clinical notes [15], provision of training, consultations, and treatments, etc.
- **Neural Networks** are highly useful in analyzing, modeling, and interpreting healthcare data. Applications of artificial neural networks (ANN) to healthcare include clinical diagnosis, prediction of cancer, medical analysis, and drug development.
- **Pattern Recognition** methods help to enhance the clinical decision support system by focusing mainly on the patient's condition, based on symptoms and demographic information. They also help to improve public health screening in healthcare systems [16].
- **Spatial Analysis** plays a vital role in the effective use of geographic information systems (GIS) in healthcare, to facilitate health data exploration, data modeling, and hypothesis testing. Spatial analysis applications include direct patient care, epidemic disease prevention and intervention, assistance with strategic planning, etc.

1.5 APPLICATIONS OF DATA ANALYTICS IN HEALTHCARE

Big data and analytics have various applications in healthcare. With the digitization of patient records, medical history, X-rays, etc., there is a tremendous opportunity for data analytics to uncover useful information. Along with the volume of data is the sheer variety of data generated from different sources at different time points. The volume of and variation in data generated by this sector are what makes it a topic of great interest for data analysts. Conventional computing mechanisms and systems fail to provide real-time monitoring and preventive plans for patients as well as for the doctor. Hence, there is a need for smart strategies that decipher the incoming data to uncover trends and anomalies and which give recommendations for patients, helping doctors in their practice.

Some applications of data analytics in healthcare are as follows.

- **Image-Based Analytics**
  Image-based datasets are a common source of information and are primarily used by doctors for internal imaging. X-rays, mammography, CT, positron emission tomography–CT (PET–CT), ultrasound, and MRI are some of the imaging technologies commonly used for diagnostics [5, 17]. Many organizations and medical institutions release open datasets in a hope to foster research activities. Neuro-images and MRI of the brain are widely used for detecting tumors, an anomaly where the cells are enlarged and form solid neoplasms. Early-stage detection is necessary for effective treatment and recovery, and for decreasing the risk of mortality [18–20]. X-rays are carried out for detecting fractures, pneumonia, cancer, etc. A large amount of research has also been carried out for analyzing X-rays and CT scans to promote early detection of the novel coronavirus COVID-19, with minimal human intervention and interaction [21, 22].

  With the growing number of medical records, the reliance on computer-aided diagnosis and analysis is increasing [8]. High-performance computing and advanced analytical methods, like ML and optimization techniques, are aiming to minimize predictive errors. Countries with low doctor-to-patient ratios can benefit greatly from such machine-aided diagnostics.

  A significant challenge associated with image-based analytics is the amount of data generated. Images are space intensive. A single X-ray can take up several megabytes (MBs) of storage space. The quality of an image plays a crucial role in correct diagnosis and hence must not be compromised. What adds to the processing challenges is that the data are highly unstructured. Image processing techniques like segmentation, denoising (noise reduction), and enhancement form the pre-requisites before useful features, like color, contour, shape, pixel intensity, edges, etc., can be extracted to train models for classification and diagnosis.
- **Signal-Based Analytics**
  In a world full of wearable sensors, time-based signals are being generated at a frequency greater than that at which they can be processed. Wearable devices, like smart watches, smart rings, and fitness trackers, continuously track heart rate, blood pressure, sleep patterns, calories burned, etc. Apart from personal devices and gadgets, time-stream data are being generated by electrocardiograms (ECG), ventilators, electroneurograms (ENG), electroencephalograms (EEG), phonocardiograms (PCG), etc. [23]. Analysis of

these signals plays a significant role in deciding and prescribing medication, care regime, routine check-ups, etc. Readings of such signals can provide useful information concerning the current status of the patient.

With lifestyle changes over the past decade, heart diseases have become more common [24]. ECG signals, providing first-hand information on the well-being of the heart, can be sensed through carbon nanotubes and sent directly to doctors for real-time analysis, while the patient may still be at home [25]. Such a set-up reduces the need for patients to physically be present at the hospital or clinic to obtain a consultation. Furthermore, a live feed can also be sent to intelligent devices which have been trained to use ML models, to identify any anomaly in the signals [26]. On detection of any abnormal signs, the doctor can be contacted.

The mental state of a driver can be monitored using physiological signal analysis. This can prevent accidents which occur due to negligence or lack of attention by the driver [27]. In stressful times like today, when people are restricted by lockdown to their homes because of a global pandemic, mental health monitoring becomes more crucial than ever. Thus, detection and monitoring of neurological and mental health markers in patients can be a game changer [28].

Signals from a wide range of sensors can be generated continuously. These signals may be analyzed periodically for a routine check-up, or continuously, if the condition of a patient is critical. Nevertheless, all data still need to be saved in the patient's history, for future reference. Imagine the amount of data being generated by an individual wearing a smart watch 24/7. As with image-based analytics, such data are enormous in volume and rich in variety. Every sensor generates a different class of data. Such time-stream data needs to be placed in context in order to derive meaningful results regarding the current status of the patient. Furthermore, it adds another layer of difficulty to the analysis of signal-based data. The complexity of the problem is increased by the need for real-time stream diagnostics and analytics.

- **Clinical Diagnosis and Research**

The study of signs, symptoms, and medical history of a patient to assess underlying conditions or disease is known as a clinical diagnosis. It usually does not involve any laboratory testing but uses the prior records of the patient, along with current symptoms and characteristics, to reach a preliminary level of diagnosis. Subsequently, additional tests can be ordered by the doctor to pinpoint the exact extent of the disease or to uncover anomalies and false symptoms. While a doctor uses years of experience to reach this first level of diagnosis, it is a time-consuming process to go through the medical history of the patient and past treatments to which the patient has been subjected.

With the digitization of data, the entire medical history of every patient is stored as an EHR. Information like clinical notes, prescriptions, administrative data, laboratory test results, medical imaging data, the patient's personal data, etc., can all be easily retrieved and stored permanently because of ever-decreasing hardware cost [17, 29].

Data analytics can be helpful not only in the detection of disease at the earliest stage, but can also attain high levels of accuracy, predicting the timeline and disease development trajectory. Furthermore, data analytics can bring to the doctor's notice whether there is any change in the vital signs, indicating a deviation from a healthy state. It can provide transparency to patients and provide them with a more personalized experience of the entire healthcare management system [8, 30]. Healthcare organizations also benefit from data analytics as they help to provide cost-effective care and personalized predictions for each patient. Clinical data can also be useful for research purposes, such as the demographics of patients with a particular condition, predicting the sales of drugs and their profitability, identifying drug competitors, usage patterns of drugs, and effective drug design, uncovering inter-drug associations, etc. [8].

Clinical decision support systems (CDSS) [5, 23, 31] may therefore be an all-round solution for automated clinical diagnostics and research. But achieving systems with high levels of accuracy and efficiency is a big challenge. The data gathered from different sources in multiple formats over time adds volume, variety, and velocity to the data. Systems with high computational power to deal with structured, semi-structured, and unstructured data need to be implemented. Furthermore, handwritten clinical notes, prescriptions, and medical journals need advanced ML algorithms with concepts of NLP. Thus, the heterogeneity of clinical data remains currently the most significant challenge and an open research area.

- **Disease Transmission and Prevention**

  Some diseases are infectious and can be spread *via* direct or indirect contact with the infected person or carrier. To prevent the outbreak of such conditions, it becomes critical to study the means of transmission and to predict the spread of the disease to develop better mitigation plans and improved disease management strategies [32]. With the help of data analytics, mathematical and stochastic models can be generated to predict the outreach of the disease and to estimate its impact.

  Many researchers have studied the transmission of the novel coronavirus that emerged in Wuhan in late 2019 and which has spread throughout the world. It has been declared by the World Health Organization (WHO) to be the worst epidemic in the past two decades [33]. Based on early available data, symptoms, numbers of positive cases, and international travel history, a predictive model was developed by researchers to identify the extent of transmission and the risk it posed to human life [34]. Many government organizations have also funded projects to research the prevention and preparedness of individual countries. Massive amounts of data gathered globally have been used to develop preventive measures to stop or minimize further spread of the virus. Many countries opted for complete lockdown, halting businesses, closing schools and universities, banning travel, etc. Medical equipment, ventilators, masks, personal protective equipment (PPE), and sanitizers have been mass produced to deal with the prevailing situation. Predictive analyses for other diseases, like HIV/AIDS, have previously been carried out to enable early detection and treatment [35]. Transmission of all chronic illnesses could be prevented if the necessary measures are undertaken, although this would be costly [36].

Data play an essential role in understanding the transmission model of communicable diseases. The success of any disease prevention model depends largely on the timely identification of the pathogen. Hence, accurate classification, clustering, and associative models need to be developed, which can help physicians to prescribe medication and other treatments [36].

- **Precision Medicine**
  With an increase in the volume of data being generated on a daily basis from sensors, implants, EHR, clinical practices, etc., the potential to foster healthcare and medical functionalities has increased dramatically. Population health management, prevention of disease transmission, and CDSS, have all become possible. Ever-increasing sizes of EHR, and their dimensionality and variety, along with the incorporation of behavioral, social, and omics data, have given data analysts the power to propose and develop models for personalized patient care and precision medicine. The layered architecture of such models, working on different aspects of healthcare, has facilitated the development of such comprehensive and detailed healthcare solutions. By incorporating data from varied sources, and identifying relationships and patterns of interest, the entire healthcare solution framework has migrated from a disease-centric view to a patient-centric view. Diverse formats of data, the generation of bulk volumes of data, and the inherent uncertainty associated with sources of big data complicate the task of data curation. Transformation of raw data into useful facts and information is vital for these healthcare framework solutions to achieve their intended goals [8].

  Precision medicine promises better healthcare delivery by improving prognosis, diagnosis, and treatments being given to patients. Improving the quality of clinical practices translates directly to personalized healthcare routines for each individual, optimized for their direct benefit [37].

  Some challenges identified for precision medicine involve upgrading of ongoing clinical practices to incorporate newer, disruptive technologies in the field of big data, the ability to handle large volumes of data generated from different data sources and their analysis to achieve meaningful results [38]. High-scale cognitive computing, using advanced ML models, can drive data-driven analysis of biomedical big data [39].
- **Health Insurance**
  Insurance agents and companies today enjoy an extensive database of customers belonging to different demographics. The application of data analytics to this large dataset can help the agents to identify patterns, clusters, and typical human behavior. With insurance models and schemes moving

online, consumers expect transparency and an accurate breakdown of the money they are spending on health insurance. Thus, the agencies cannot add some hidden costs to make additional profits. Competition among agencies is stiff and, hence, attracting customers with lucrative benefits is how these business agencies function [40].

Insurers today need to focus on individual customizable plans for individuals, instead of targeting groups of people. Predictive analysis is an essential branch of analytics, which is widely used in such scenarios. It identifies the appropriate coverage for the needs of people and further customizes plans based on individual inputs. Smart models deployed online first take the necessary information from users before developing proposals for each individual. This personal experience and the tailor-made environment are valuable for customers.

The insurance costs can be intelligently calculated based on the current health and medical history of an applicant. For instance, with the growing obesity rate, the insurers may calculate the potential risk associated with an individual applicant and modify the premium for health insurance accordingly. Furthermore, as obesity increases the risk of heart failure, preliminary tests may be advised before the policy can be issued [41].

Thus, health insurance companies use big data analytics to deliver a personalized experience to each individual, to predict the occurrence of fraud before it is realized, and to fast-track claims by predictive analytics.

- **Service Delivery Systems**

Today, healthcare providers across the globe are facing competition and understand the need to deliver the highest quality of services to their customers. Companies work on low profit margins and hence need to optimize their service delivery systems to minimize the cost of providing the service.

The easiest way to do this is by automating the healthcare services and reducing human intervention with respect to suggesting remedial actions, preventive measures, prescriptions, diagnostics, etc. The majority of the costs incurred by such organizations are on high-quality medical equipment that needs to be upgraded over time, to keep pace with global technological advances. The research and development cost for new medicines is also subject to extensive clinical trials before they can be rolled out to the public for consumption, leading to high costs [42].

Furthermore, medical devices, equipment, and sensors continuously generate large volumes of data which can be used for forecasting and decision making. Hence, data analytics can serve as a powerful tool for improving healthcare delivery systems.

A service delivery network usually consists of multiple organizations working together collaboratively to provide an overall healthcare service delivery system. These organizations are closely connected with one another and share data among themselves to make smart, informed decisions. Players involved in such networks may include sales representatives, doctors, physicians, insurance personnel, laboratory technicians, hospital staff, equipment vendors, etc. They continuously explore ways to understand data being generated from different sources and to understand the impact such data have on their policies and services. This co-development and deployment environment aims to reduce the cost of healthcare services [43, 44].

1.6 CHALLENGES ASSOCIATED WITH HEALTHCARE DATA

With the use of big data analytics, the healthcare industry has improved in many aspects, including operational efficiency in healthcare management, reduction in healthcare costs, improved drug discovery, higher-quality healthcare services, personalized patient care, effective treatments, and improved clinical outcomes. Despite the benefits, however, big data introduces several challenges.

- One of the foremost challenges facing the healthcare industry nowadays involves capturing health-related data. Due to the extensive use of IoT devices and wearables in the smart healthcare system, which constantly generate massive volumes of streaming data, the capture and processing of data has become a major challenge. Lack of efficient governance practices for data sources is another challenge for the capture of data from heterogeneous data sources, a problem which can lead to inaccurate data.
- Due to the exponential volume growth of streaming health data, data storage has become a primary challenge for the healthcare industry. Most health organizations prefer in-house data storage so that they can have control over data security and data access, but, with the rapid growth of health data, in-house data storage infrastructures become difficult to scale up as maintenance and scale-up costs are high.
- The privacy and confidentiality of medical data, in the form of a patient's health data, are of utmost importance in healthcare. Data sharing among the various practitioners in a healthcare system escalates the need for privacy, while informed consent issues are another challenge faced by data analytics in healthcare.
- Since healthcare data are generated from multiple sources, such as medical images, wearable sensors, and EHRs, there are no fixed unified standards for these data, leading to difficulties in consolidating and processing of the data. Therefore, the lack of data protocols and standards are one of the governance challenges facing healthcare data analytics.
- As digital healthcare has no geopolitical boundaries, health services are available across international and national borders. It is difficult to form uniform legislation because medical licenses, privacy of patient data, and the advertisement and marketing of healthcare services may vary between countries, representing major challenges for the healthcare industry.
- Data analytics in healthcare also pose new ethical and legal challenges, including personal autonomy, risk of compromising the patient's privacy, need for informed consent, trust and transparency when using biomedical big data.

1.7 CONCLUSION

Data analytics with big data in healthcare is still at the developing stage and advances in tools and techniques will improve and their applications will expand. In addition, establishing proper standards and governance of data, ensuring data privacy and security, and updating the healthcare systems continuously are some of the challenges faced by the healthcare industry. Improving communication and data sharing among related sectors in healthcare would increase the overall efficiency by providing value-added services, with minimal additional costs incurred.

REFERENCES

1. A. Belle, R. Thiagarajan, S. M. R. Soroushmehr, F. Navidi, D. A. Beard, K. Najarian. Big data analytics in healthcare. *Biomedical Research International*, 2015, 16 pages, 2015. http://dx.doi.org/10.1155/2015/370194.
2. A. Kankanhalli, J. Hahn, S. Tan, G. Gao. Big data and analytics in healthcare: Introduction to the special section. *Information Systems Frontiers*, 18, 233–235, 2016.
3. M. J. Ward, K. A. Marsolo, C. M. Froehle. Applications of business analytics in healthcare. *Business Horizons*, 57, 571–582, 2014.
4. S. Kumar, M. Singh. Big data analytics for healthcare industry: Impact, applications, and tools. *IEEE*, 2 (1), 48–57, 2019.
5. N. Mehta, A. Pandit. Concurrence of big data analytics and healthcare: A systematic review. *International Journal of Medical Informatics*, 114, 57–65, 2018.
6. Y. Wang, N. Hajli. Exploring the path to big data analytics success in healthcare. *Journal of Business Research*, 70, 287–299, 2017.
7. N. El aboudi, L. Benhlima. Big data management for healthcare systems: Architecture, requirements, and implementation, Hindwai. *Advances in Bioinformatics*, 2018. https://doi.org/10.1155/2018/4059018.
8. V. Palanisamy, R. Thirunavukarasu. Implications of big data analytics in developing healthcare frameworks–A review. *Journal of King Saud University-Computer and Information Sciences*, 31(4), 415–425, 2019.
9. S. Melnik, A. Gubarev, J. J. Long, G. Romer, S. Shivakumar, M. Tolton, T. Vassilakis. Dremel: Interactive analysis of web-scale datasets. In: *36th International Conference*, 2010.
10. C. E. Seminario, D. C. Wilson. Case study evaluation of Mahout as a recommender platform. In: *6th ACM Conference on Recommender Engines (RecSys 2012)*, pp. 45–50, 2012.
11. https://www.softwaretestinghelp.com/big-data-tools/.
12. https://www.predictiveanalyticstoday.com/top-open-source-commercial-stream-analytics-platforms/.
13. https://www.yourtechdiet.com/blogs/impact-data-visualization-healthcare/.
14. N. Downing, A. Cloninger, A. Venkatesh, A. Hsieh, E. Drye, R. Coifman, et al. Describing the performance of U.S. hospitals by applying big data analytics. *PLoS One* 12(6), e0179603, 2017.
15. A. Khalifa, S. Meystre. Adapting existing natural language processing resources for cardiovascular risk factors identification in clinical notes. *Journal of Biomedical Informatics*, 58, S128–S132, 2015.
16. D. D. Luxton, J. D. June, A. Sano, T. Bickmore. Intelligent mobile, wearable, and ambient technologies for behavioral health care. *Artificial Intelligence in Behavioral and Mental Health Care*, Elsevier, 137, 2015.

17. B. Ristevski, M. Chen. Big data analytics in medicine and healthcare. *Journal of Integrative Bioinformatics*, 15 (3), 1–5, 2018.
18. A. R. Kavitha, C. Chellamuthu. Brain tumour detection using self-adaptive learning PSO-based feature selection algorithm in MRI images. *International Journal of Business Intelligence and Data Mining*, 15 (1), 2019.
19. S. Tchoketch Kebir, S. Mekaoui, M. Bouhedda. A fully automatic methodology for MRI brain tumour detection and segmentation. *The Imaging Science Journal*, 67(1), 42–62, 2019.
20. T. V. N. Rao, H. Katukam, D. Guvva. Early brain tumour detection in MRI using enhanced segmentation approach. image, 8, 9, 2019.
21. L. Brunese, F. Mercaldo, A. Reginelli, A. Santone. Explainable deep learning for pulmonary disease and coronavirus COVID-19 detection from X-rays. *Computer Methods and Programs in Biomedicine*, 196, 105608, 2020.
22. A. Jacobi, M. Chung, A. Bernheim, C. Eber. Portable chest X-ray in coronavirus disease-19 (COVID-19): A pictorial review. *Clinical Imaging*, 64 (April), 35–42, 2020.
23. C. K. Reddy, C. C. Aggarwal. An introduction to healthcare data analytics. In *Healthcare Data Analytics*, 2015, pp. 1–18.
24. S. Dalal, V. P. Vishwakarma. GA-based KELM optimization for ECG classification. *Procedia Computer Science*, 167, (2019), 580–588, 2020.
25. M. Bansal, B. Gandhi. IoT & Big Data in Smart Healthcare (ECG Monitoring). In *2019 International Conference on Machine Learning, Big Data, Cloud and Parallel Computing (COMITCon)*, 390–396, IEEE, 2019, February.
26. S. Dalal, V. P. Vishwakarma, V. Sisaudia. ECG classification using Kernel extreme learning machine. In *2nd IEEE International Conference on Power Electronics, Intelligent Control and Energy systems (ICPEICES-2018)*, pp. 988–992, 2018.
27. Barua, S., Ahmed, M. U., & Begum, S. Distributed multivariate physiological signal analytics for drivers' mental state monitoring. In *International Conference on IoT Technologies for HealthCare*, pp. 26–33, Springer, Cham, 2017, October.
28. M. Neumann, O. Roesler, D. Suendermann-oeft, V. Ramanarayanan. On the utility of audiovisual dialog technologies and signal analytics for real-time remote monitoring of depression biomarkers. In *Proceedings of First Workshop on Natural Language Processing for Medical Conversations*, pp. 47–52, 2020.
29. Belle, A., Thiagarajan, R., Soroushmehr, S. M., Navidi, F., Beard, D. A., & Najarian, K. Big data analytics in healthcare. *BioMed Research International*, 2015, 2015.
30. Wang, Y., & Hajli, N. Exploring the path to big data analytics success in healthcare. *Journal of Business Research*, 70, 287–299, 2017.
31. Shafqat, S., Kishwer, S., Rasool, R. U., Qadir, J., Amjad, T., & Ahmad, H. F. Big data analytics enhanced healthcare systems: A review. *The Journal of Supercomputing*, 76(3), 1754–1799, 2020.
32. Wong, Z. S., Zhou, J., & Zhang, Q. Artificial intelligence for infectious disease big data analytics. *Infection, Disease & Health*, 24(1), 44–48, 2019.
33. Koubâa, A. Understanding the covid19 outbreak: A comparative data analytics and study. arXiv preprint arXiv:2003.14150, 2020.
34. Kucharski, A. J., Russell, T. W., Diamond, C., Liu, Y., Edmunds, J., Funk, S., ... & Flasche, S. Early dynamics of transmission and control of COVID-19: A mathematical modelling study. *The Lancet Infectious Diseases*, 20(5), 553–558, 2020.
35. Das, N., Das, L., Rautaray, S. S., & Pandey, M. Detection and prevention of HIV aids using big data tool. In *2018 3rd International Conference for Convergence in Technology (I2CT)*, pp. 1–5. IEEE, 2018, April.
36. Razzak, M. I., Imran, M., & Xu, G. Big data analytics for preventive medicine. *Neural Computing and Applications*, 32(9), 4417–4451, 2020.

37. Panayides, A. S., Pattichis, M. S., Leandrou, S., Pitris, C., Constantinidou, A., & Pattichis, C. S. Radiogenomics for precision medicine with a big data analytics perspective. *IEEE Journal of Biomedical and Health Informatics*, 23(5), 2063–2079, 2018.
38. Hulsen, T., Jamuar, S. S., Moody, A. R., Karnes, J. H., Varga, O., Hedensted, S., ... & McKinney, E. F. From big data to precision medicine. *Frontiers in Medicine*, 6, 34, 2019.
39. D. Cirillo, A. Valencia. Big data analytics for personalized medicine. *Current Opinion in Biotechnology*, 58, 161–167, 2019.
40. Gupta, S., & Tripathi, P. An emerging trend of big data analytics with health insurance in India. In *2016 International Conference on Innovation and Challenges in Cyber Security (ICICCS-INBUSH)*, pp. 64–69, IEEE, 2016, February.
41. Revels, S., Kumar, S. A., & Ben-Assuli, O. Predicting obesity rate and obesity-related healthcare costs using data analytics. *Health Policy and Technology*, 6(2), 198–207, 2017.
42. Alotaibi, S., Mehmood, R., & Katib, I. The role of big data and twitter data analytics in healthcare supply chain management. In *Smart Infrastructure and Applications*, pp. 267–279, Springer, Cham, 2020.
43. M. A. Pikkarainen. Data as a driver for shaping the practices of a preventive healthcare service delivery network. *Journal of Innovation Management*, 1, 55–79, 2018.
44. M. Usak, M. Kubiatko, M. Salman. Health care service delivery based on the Internet of things: A systematic and comprehensive study. *International Journal of Communication Systems*, 33, 1–17, 2019.

2 Systematic View and Impact of Machine Learning in Healthcare Systems

Neethu Narayanan, Arjun K. P., Sreenarayanan N. M., and Deepa C. M.

CONTENTS

2.1 INTRODUCTION

Artificial intelligence (AI) methods [1–3] have become among the main discussion points inside the healthcare system, especially for clinical choice emotionally supportive networks, which are regularly used to help doctors to make more precise analyses. In either case, applying these AI methods to clinical support systems undoubtedly highlights the absence of models suitable for use. Consequently, recent investigations have zeroed in on assessing distinctive AI classifiers, with the aim of identifying the most fitting classifier to be used for specific dynamic issues. Most of these investigations have used a solitary dataset inside a specific clinical-related characterization space.

Assessing AI classifiers with one example of information gives the impression of being sub-optimal, perhaps because it is not reflecting the classifiers' capacities or their personal conduct standards under different conditions. In this current investigation, five notable regulated AI classifiers will be analyzed, using five distinct global datasets with a range of parameters. The fundamental aim [4] was to represent not just the effect of the dataset's volume and qualities on the assessment, but also (and all the more significantly) to present the classifiers' strengths and weaknesses under specific conditions, to possibly provide direction or guidance to help health specialists to decide the most appropriate classifier to address a specific, clinically related dynamic issue.

Harmful growths and tumors have been shown to be a condition consisting of different subtypes [5]. The early analysis and diagnosis of a harmful tumor type has become a requirement in disease research, as it can target the resulting medicine-based management of patients. The importance of classifying patients with harmful cancers into high or generally safe classes has driven many research groups, from the study of how life and medicine work together, as well as from the bioinformatics field, to think about the use of AI. In this manner [6], these methods have been used as a means by which to show the therapy and responses of dangerous cancer conditions. Furthermore, the ability of machine learning (ML) to identify key highlights from complex datasets reveals their importance.

A number of these procedures, including Artificial Neural Networks (ANNs), Bayesian Networks (BNs), Support Vector Machines (SVMs) and Decision Trees (DTs), has been widely applied in cancer research for the improvement of smart models. Even though the use of ML can improve our understanding of cancer development, an appropriate level of approval is required for using these methodologies. There are many different types of ML approaches used in the display of cancer development. The smart models examined here depend on different managed ML methods and on different information highlights and information tests. Given the developing pattern of the use of ML in cancer research, we will describe here the latest strategies that use these methods to reveal the risk of cancers and the impact on the patient.

2.2 APPLIED ML IN HEALTH CARE

As of late, there has been a dramatic increase in the use of AI methods within the medical care frameworks to analyze, predict, and classify clinical information. Uses of AI can improve the utilization of information, and, in this manner, to improve the nature of medical care being administered to the patient [7, 8]. The idea of AI resembles a computer program that takes in and collates information from past experience or potentially through distinguishing the significant highlights of a dataset provided, so as to make predictions about other data that were not included in the original dataset [9]. Analysis of AI procedures has increased its value for clinical choice supportive networks, which are broadly used in the medical services sectors to assist doctors in making more precise diagnoses, decreasing the frequency of clinical errors, and improving patient security. ML classifiers are accepted to be the backbone for most of the complex clinical system, where ML procedures are introduced as one of the primary segments of the data handling suggested, the insight module being a fundamental part of clinical support, empowering such frameworks to learn after a period of time. In addition, a study, directed by Foster and colleagues, demonstrated that receiving support from ML procedures would help deal with decision making. Similarly, selecting a specific ML classifier to be executed to provide clinical support is a basic advance, which, in all probability, would overcome a lack of measures with application of ML classifiers. Several recent studies e.g., [10] have analyzed and assessed diverse directed ML classifiers, using a particular dataset. The assessment of ML classifiers by using a single example of information is undoubtedly sub-optimal, maybe not reflecting the classifiers' abilities or their behavior under different conditions. Along these lines, this current study intends to evaluate and assess five notable regulated ML classifiers by utilizing five different datasets with a range of characteristics, including size. The ML classifiers that have been focused on are multilayer perceptron, naïve bayes, stochastic gradient descent support vector machine and *k*-nearest neighbor. This should show not just the effect of the dataset volume and characteristics on the assessment, but would also (and all the more significantly) reflect the classifiers' abilities and deficiencies under certain conditions, which conceivably will provide a direction or directions to help health analysts to select the most appropriate classifier with which to address a specific, clinically related dynamic issue.

This study is meant to:

a) Survey and talk about five mainstream administered AI classifiers;
b) Survey a range of related examinations that have zeroed in on evaluating and assessing ML classifiers; and
c) Assess these five well-known classifiers under various conditions by utilizing five datasets with various number of samples and types of credits.

2.2.1 ML-Assisted Radiology and Pathology

Nowadays, there are copious volumes of electronically stored clinical imaging data, and deep learning (DL) calculations can be taken care of with this sort of dataset, to distinguish and find examples and anomalies. Machines and calculations can decipher the imaging information much like an exceptionally prepared radiologist could, identifying dubious spots on the skin, injuries, tumors and brain drains. The use of AI to help radiologists, therefore, is likely to expand exponentially.

This methodology tackles a basic issue in the medical services area on the grounds that, throughout the world, experienced radiologists are becoming more difficult to access. As a rule, such talented specialists are under huge strain because of the storm of computerized clinical information. According to the article [11], a radiologist would need to interpret the results from one image each 3–4 seconds in order to fulfill the demand.

2.2.1.1 ML For Increased Imaging Precision in Radiology

Recognizing uncommon or difficult-to-analyze conditions frequently relies upon distinguishing the so-called 'edge cases'. As this sort of ML framework is based on huge datasets, containing crude images of these conditions, ML is frequently more reliable than human operators for this kind of identification.

ML algorithms are relied upon to improve the nature of image analysis for patient diagnosis within public medical services systems. Such situations could represent the greatest impact of AI applications in health care, as it can conceivably change the lives of billions of individuals around the globe.

One astounding development is Microsoft's Project InnerEye, which utilizes ML techniques to characterize tumors, utilizing 3D radiological images. It can help to increase the precision of medical procedures, by determining tumor shape for the design of radiotherapy treatments.

2.2.2 Identification of Rare Diseases

More than 6,000 uncommon diseases are described by a wide variety of symptoms, that vary from patient to patient. Generally, basic characteristics can conceal hidden uncommon conditions, prompting misdiagnosis and delaying treatment [12]. The patient's personal wellbeing is negatively affected by the symptoms of the condition. The lack of a consistent effective treatment adds to the negative impacts suffered by patients and their families.

2.2.2.1 Regular Challenges

The absence of information and high-quality data on the ailment regularly brings about a delay in diagnosis. Furthermore, the application of appropriate therapies with which to treat the condition is often inconsistent, leading to substantial social and budgetary weights on patients.

Because of the wide variety of issues and symptoms which can mask diagnosis of hidden, uncommon diseases, initial misdiagnosis is quite common, as symptoms and side effects can vary from patient to patient. Because of the range of uncommon diseases, the results of research need to be conveyed globally to guarantee that specialists, analysts, and clinicians are all informed, so that patients can benefit by the pooling of assets and information among countries. The International Rare Disease Research Consortium and the Horizon 2020 EU Framework Programme for Research and Innovation support worldwide research associated with rare diseases.

Rare Disease Day, which is observed on the last day of February each year, brings issues to light for the 300 million individuals living with uncommon diseases around the globe, as well as their families and carers. The goal of the Rare Disease Day crusade is to accomplish fair access to diagnosis, treatment, health, and social consideration for all individuals influenced by a rare disease. Significant advances are being achieved with worldwide health and wellbeing promotion attempts, as part of the United Nation Sustainable Development Goals to advocate for fair wellbeing frameworks that address the issues of individuals affected by rare diseases, so that no-one is left behind. Rare Disease Day is the open-door advocate for uncommon illnesses at local, regional, and worldwide levels, as we aim to achieve a more inclusive society [13].

2.2.3 ML in Mental Health Care

The effect of ML in psychological wellness will be included in the plan of frameworks which employ ML, which spurs us to analyze ongoing studies with respect to data processing and healthcare informatics (HCI) in this discipline. Supplementing research findings from clinical science and clinical brain science, our current chapter presents a systematic survey of the computing literature to develop a more profound understanding of the ebb and flow of ML applications for emotional wellness from a HCI and processing science viewpoint. Our work expands on an recent survey [14], which planned advances for support of wellbeing, as revealed by HCI, and identified that most development has happened in the areas of computerized diagnosis.

As scientists, who are effectively working at the crossroads of HCI, ML, and emotional wellbeing, we are excited by the potential benefits that ML strategies could bring to psychological wellbeing. All the while, from the start of the audit, we were additionally mindful that the improvement of successful and implementable ML frameworks is bound up with a variety of intricate, interlaced socio-specialized problems. As a result, our survey is probably colored by both our mindful good faith that ML approaches can be conveniently and effectively applied to this discipline, and that a solid human-focused point of view on innovation advances is just as important in making dependable AI applications that look to improve cultural results. Accordingly, we have taken, on occasion, a marginally more basic view of research that proposes mediations that could bring about significant improvements, although remaining exclusively focused on technical advances. We welcome developments of the network, in mix with other approaches, to achieve fruitful utilization of ML in addressing psychological wellbeing.

These problems include the enormous datasets generated, which are illustrative of the variation in the populace, and the need to access such datasets to develop more powerful and valuable ML models. Psychological wellbeing, specifically, influences an expansive range of individuals – crossing various socioeconomic groupings (age, sexual orientation, nationality), geographic areas, and financial statuses – which require the inclusion of a wide range of individuals for such variation to be reflected in the dataset to relieve risks of bias. In addition, information sorting is expensive and problematic, especially where data are profoundly personal, because of the shame that is frequently associated with psychological issues. This main issue of whether individuals should consent to the inclusion of their personal data in ML datasets.

2.3 MAJOR APPLICATIONS

2.3.1 Personalized Medicine

Customized or personalized medicine involves an extensive automated information base that allows the design of patient profiles. For a patient who presents at a center, a range of quantitative clinical and genomic assessments are derived, leading to the development of the patient's profile vector. Every element of this vector relates to a biomarker (or symptom) important to carry out diagnosis. For example, a profile vector can relate to quality parameters estimated either by DNA microarrays or genomic parameters. For such a situation, part of the vector relates to the quality articulation estimate of an individual parameter, such as the abundance of a specific mRNA or even of non-coding RNA, if RNA-Seq data are available.

Interestingly, such profile vectors contain useful estimations of the changes in phenotype on a genomic scale; such associations, developed by non-ML methods, are usually inadequate.

In straightforward terms, such a study consists of a correlation between the profile vector of the patient with an information database that gives comparable profile vectors to a population of patients along with clinical result factors, e.g., endurance time or response to prescriptions. The association between reference profiles and clinical result factors builds up a system that takes into consideration the profile of the patient. Basically, this implies the best coordination of the profile of a patient, taking into consideration proposals for the detection, diagnosis, and therapy trajectory of conditions of the patient. In a perfect world, these proposals would be real choices in an automated technique. This automation step would likewise be vital for identifying the customized medication from the available conventional medication.

The exhaustiveness of the information base and the approach taken are significant in light of the fact that, as described above for the "phenotype classification", an absence of reference information causes some illness phenotypes to be overlooked. This infers that customized medication is necessarily a network system, since singular approaches cannot cover the total of medication available. At long last, we can take note directly affect the profiles of the patient, using the information database and a metaphysical approach in an interconnected way.

2.3.1.1 Viable Personalized Medicine

The above portrayal of customized medication from an AI point of view can be viewed as optimistic. The purpose behind this is twofold. To start with, the breadth of the information base available just for oncology is enormous, counting the many subtypes of tumors and the known treatments for each of these. Right now, we are far away from this stage, with the information available being seriously deficient. On the other hand, the subsequent information base for oncology that can be developed, based on our present data, can be very valuable. The subsequent explanation concerns the automation of the entire cycle. The acknowledgment of this progression in any thorough way is likewise now distant [15].

2.3.2 Autonomous Robotic Surgery

Robots can provide interesting and valuable help to human surgeons, upgrading the capacity to detect and explore in a technique, making exact and negligibly intrusive entry points, and causing less pain, with faster wound healing. There are genuinely exciting opportunities for the utilization of AI with such computerized medical procedure robots. A product-driven coordinated effort of robots, guided by an information-driven strategy and directed by medical procedure narratives (performed by both machines and people) and their results (good or not), AI has produced computer-generated reality space for constant training and direction, with the potential for telemedicine and distant medical procedures for moderately straightforward operations

2.4 ML in Cancer Diagnostics

Over the past number of years, a development associated with cancer research has been performed [16, 17]. In the beginning, the goal was to detect types of disease before they caused visible symptoms. Since then, new procedures have been developed for the early diagnosis of the treatment of harmful growths. With the appearance of new inventions in the field of medicine, much information on cancers and the medicine-based examination network has been collected. The exact diagnosis of a disease is one of the most testing assignments for doctors. Therefore, ML has become a mainstream mechanical device for medicine-based analysts. Using complex datasets, these procedures can find and recognize samples and associations between them, while they can also successfully predict the future trajectory of a harmful cancer type.

Given the importance of customized medicine and the developing pattern of the use of ML procedures, we present here a survey of studies that use the ML approach to diagnosis, with respect to the diagnosis and treatment of malignant cancers. In these studies, prediction-related and smart highlights of a clearly stated therapy are considered or are incorporated to control treatment for individual cancer patients. Furthermore, we discuss the types of ML being used, the kinds of information they generate, and the general presentation of each proposed treatment plan, while also examining their strengths and weaknesses.

The proposed work incorporates a mix of blended information, for example, medicine-based. In any case, an issue that we saw in several examples is the external assessment with respect to the presentation of such ML models. The quality of disease prediction has improved by 15%–20% in recent years, with the use of ML methods.

A few studies have been described in the literature, which depend on alternative ways of providing support to the early decision making in the diagnosis of cancers [18]. In particular, these studies describe approaches using microRNAs (miRNAs), which have proved to hold promise for recognition and identification of cancers. Patients subjected to this approach experience problems of low effectiveness with respect to their use at the early stages of cancer development, and difficulties with distinguishing (because of race, religion, etc.) benign from malignant tumors.

Different points of view, with respect to the expectation of cancer diagnosis, depending on clear, quality markers, are examined. These studies list the potential and the limitations of microarrays for the forecast of harmful growths of tumor. Even though quality markers could, theoretically, improve our ability to detect disease in patients, there has been little benefit from their use in clinical centers. Before clear decisions can be made in medicine-based practice, larger information datasets and more acceptable assessments are needed.

2.4.1 ML and Cancer Imaging

The wish to improve the ability to detect, diagnose, and treat diseases keeps on driving different groups to work on new strategies, including AI. With the expanding interest from healthcare managers, the huge amounts of information created day by day, the improvement of medicine-based work processes has become fundamental. AI excels at perceiving complex examples in images and, in this manner, offers the chance to change image interpretation from a simple and subjective task to one that is countable and easily reproducible. Also, AI may measure data from images that are not capable of being carried out by a human operator because of previous knowledge and in that way add to medicine-based patterns of behavior. Computer-based intelligence can give power to the aggregation of different information streams into an indicative approach, going through radiographic images, electronic health records, and informal organizations. Within disease imaging, AI is very effective at performing three medicine-based tasks: identification, representation, and diagnosis of tumors. Location suggests restriction to objects of particular interest to radiographers [19]. Intelligence-based location can be used to reduce oversights. Described inside a sample admission setting, areas with doubtful imaging qualities are featured and introduced to the careful reader. Computer-aided detection (CADe) has been used to help distinguish mixed tumors in low-dose computed tomography (CT) to identify brain tumors in magnetic resonance images (MRIs) to improve contrast resolution compared with X-rays, to identify microcalcification groups in medical breast examinations as an indicator of early breast cancer, among others. Images can capture the division, end result and arrangement of tumors, as well as predicting the trajectory of a given disease, as well as identifying appropriate ways of treating the disease. Division describes the degree of difference between the diseased and typical version. This can extend from basic two-dimensional (2D) estimates of the maximal in-plane tumor measurements, to include volumetric divisions, in which the whole tumor and associated tissues are surveyed. Such data can be used in resulting assignments, as well as measurement during radiation treatment. In current medicine-based practice, tumors are commonly physically described, with related restrictions, including inter rater bias and lack of reproducibility even among experts, involving expenditure of time and effort. In spite of the mostly used as the reason for making a decision about the computerized division calculations, it can possibly ignore subclinical illness. Computer-based intelligence can possibly increase the effectiveness, reproducibility, and nature of tumor size estimates significantly, with robotized measurement. Ultimately, with the rapid development of estimating speed and the expanded effectiveness of AI calculations, future examination of harmful growths will not need a different division step, and whole-body imaging information could be tested by AI calculations. A whole-body approach can also permit an examination of organ structures that might not yet be obvious to the naked eye. In this way, the distinction of growths as being either benign or malignant at last brings about a visual interpretation. Human experience and ability are applied to understanding such issues. Using mathematical models, computers have helped decisions be made, based on tumor characteristic highlights, taking into account more reproducible descriptors. Careful study of lung nodules in poor-quality part CT15 lung lesions by multiparametric MRI allows anomalies to be identified.

2.4.1.1 Convolutional Neural Network (CNN) Imaging

CNN calculations are a subclass in the hierarchic phrasing that incorporates AI and deep learning [20]. Figure 2.1 presents a Venn diagram of this wording chain of command. Intelligence shows calculations that tackle issues that require human knowledge. AI is a subclass of ML, something being used. Rather, the deep learning calculations learn on their own which highlights are best for the task. Most deep learning calculations depend on neural organizations. A CNN is a subcategory of a neural network organization, that makes it clear that the sources of information are images. In the past few years, CNN has been behind the most convincing developments in the field of computer vision. Figure 2.1 shows the Venn diagram of CNN.

The CNN design contains a series of layers that change the image quantity into yield class scores in Figure 2.2. Each layer changes one volume to another through a differentiable ability. The ways in which layers are grouped together to construct a CNN are the convolutional layer, the pooling layer, the nonlinear layer, and the completely connected layer, which will be examined further below.

CNN design involves a succession of layers that change the image volume into yield class scores. Each layer changes one volume of actuations to another through a differentiable capacity. There are four fundamental kinds of layers that are joined to assemble a CNN: convolutional (Conv), pooling (Pool), nonlinear (corrected direct unit [ReLU]), and completely connected layers.

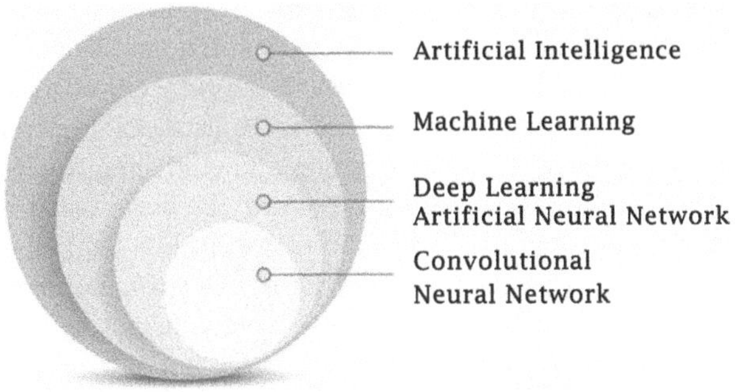

FIGURE 2.1 Venn diagram portrayal of Convolutional Neural organizations in the man-made hierarchic phrasing.

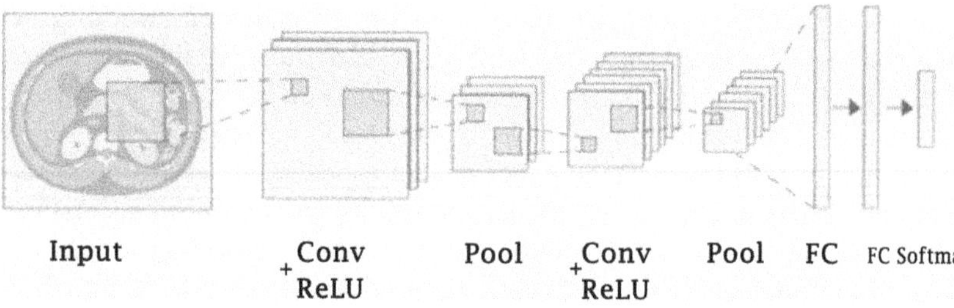

FIGURE 2.2 An ordinary Convolutional Neural Network (CNN) design for image characterization.

2.4.1.2 Radiographic Imaging

Radiography is an imaging method, using X-rays, gamma rays, or similar ionizing or non-ionizing radiation, to see the inner characteristics of an item. Uses of radiography include medicine-based radiography and mechanical radiography [21] (Figure 2.3).

Radiography is used to carefully study or treat patients by recording images of the internal structure of the body to confirm the presence or absence of sickness, unfamiliar items, and anomalies. During radiography, an X-ray passes through the body.

2.4.1.3 Digital Pathology and Image Specimens

Advanced images change with respect to their goal, shading deep thinking, and the methods by which they are stored, shared, and sent. Samples are obtained initially as organic tissues, which are then handled into tapes, inserted into paraffin squares, and afterward sliced into sections on slides. Samples do not dispense with the need for tissue-handling steps. Sadly, this requirement results in more confusion in the work process. For instance, with respect to record keeping, after having created the images, the real glass slides still need to be stored. In addition, to maintain a total informational index, any slides treated by immunohistochemistry or any other clearly stated treatment must be checked independently and added to the index of the related case. Defenders of these new slides need to make a convincing case for storage of the slides. In the following paragraphs, we present a few focal points of bringing these moderately new ways of slide preparation into the work process.

There are a few significant issues that must be carefully considered when dealing with glass slides. The most significant choice is the cautious decision about which magnification to view the slide, as it is wrong and confusing to check each slide at different magnifications. Doing so would quickly increase the size of the information for each slide and sample. It is important to pick a single magnification at which to check the slide. This decision depends on the particular application, since checking a slide at low magnification and subsequently increasing it may generate a pixelated image or an image picture that appears to be out of focus [22].

Even short-shading varieties may at times cause problems with image quality. One gathering of specialists recently considered the effect of changing the red-green-blue shading equalization on the clarity of the images. The operators found

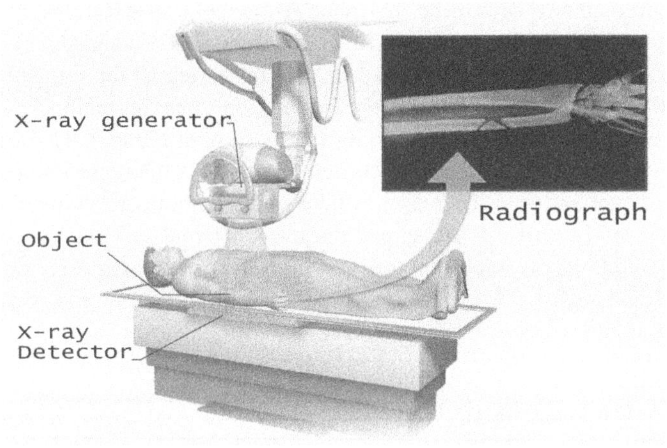

FIGURE 2.3 Radiographic imaging.

that, by changing these slides, decisions were influenced. Obviously, shadings and screenings, with the ability to precisely deliver and show tones, were needed. The advantage of using optical channels to improve the goal and shading balance to survey these images has already been pointed out. The use of programmed histology re-coloring to create top-quality slides from prepackaged stains that would not be influenced by human intervention has been shown to produce slides displaying more reliable tone, and improved division and detail, which generated better images when checked digitally.

In contrast to careful samples, the main part of the material in cytopathology exists just on the slide itself. Sometimes, there might be extra material contained inside the cell block, which can be used to represent the separation and arrangement of the cells. Anyway, there is no other tissue block from which a cytopathology slide can be copied. So, one must be careful when taking care of these samples on the grounds that there is no safeguard arrangement if the slide is lost or damaged. Another test in cytopathology samples is the importance of protecting the height, width, and depth of the cells in order to boost the value of the sample. Studies that have used just static pictures at one plane of center have pointed out significant errors between the finding created by carefully reading a filtered image in comparison with using an ordinary microscope. Reading these samples at just a number of central planes is not as effective as doing so at a large number (i.e., the majority) of the central planes.

2.4.1.4 Image Database

The ability to generate images has opened up more opportunities to store a lot of data which can be provided to different partners and teammates for meetings and examinations. Dataset models have been proposed to store these data next to any results, for explanations and divisions that may have been created throughout image analysis [23]. Data received from the images can be coordinated with data obtained from measurements to determine the shape of the cells in questions, to achieve a better understanding of the condition. Specialists have recommended making whole-slide image (WSI) forms of distributed images in the report for survey by the network, in order to confirm the data before leading their own research.

It has been a test to the way of doing things to obtain and store these images. Effective computerized imaging and communications in medicine can be achieved by storing images and data. Therefore, patient's information, orders, medicine-based data, images, and comments are spread out over different systems. A central common store would be very important to allow analysts to compare files of images from different establishments, while making it possible to make image microarrays that are comparable to physical tissue microarrays.

2.4.2 Cancer Stage Prediction

2.4.2.1 Determination of Tumor Aggression Score (TAS)

The tumor-related histopathological edges were powerful, and could be used for the prediction of the tumor phase of a malignant colon growth [24]. The tumor size was estimated after formalin fixation. The tumor size was recorded as length × width. During the evaluation, the colon was regularly cut open. The tumor 'length' shows the 'longitudinal degree' in cm, whereas the tumor 'width' represents the 'cross over degree' up-and-down to the tumor length in cm. Each of these edges was recorded. Even though the tumor size was estimated based on the tumor length and width. In this way, the most extreme size of the tumor (T_{max}) was determined using Equation (2.1) as follows:

$$T_{max} = \max(\text{tlen}, \text{twid}) \tag{2.1}$$

where tlen and twid represent the Tumrlen (tumor length) and Tumrwid (tumor width), respectively. Another interesting edge identified with the tumor, cirlnvo, normally shows the invasion of the tumor towards the divider. If the tumor is very much separated, it expands the location and the other way around. At the point when all the interesting edges were decided, Equation (2.2) was created from just those tumor-related edges/borders, as shown below:

$$TAS = T_{max} + \text{Cinvo} + T_{dif} \tag{2.2}$$

where C_{invo} and T_{dif} refer to the circumferential association and the tumor separation, respectively. Equation (2.2) was used to ML calculations from the scikit-learn software ML system in Python.

2.4.2.2 AI Analysis of ML Models

The new feature, TAS, is used as a prediction-related factor to examine its effect on the TNM classification system for malignant tumors. It was important, therefore, to evaluate the value of that quality for predicting the tumor stage and disease-free survival (DFS) time of patients with colon cancer. The information obtained contains different highlights and a huge number of events. It is hard to correspond the highlights physically and to predict ahead to the prognosis of a patient as far as tumor arrangement or DFS period goes. ML classifiers, for example, Random Forest, Support Vector Machine (SVM), Logistic Moving Backward (LR), Multilayer Perceptrons (MLP), *K*-Nearest Neighbors (KNN), and Boosting (AdaBoost) [25], are well known in medicine-based information examinations. Executing different classifiers in scikit-learn for the multi-class grouping, the plan of one-against-one was used for SVM and the plan one-vs-rest (OVR) was used for LGR and different models, which generated all the measurements used in our study. In total, 4003 samples were used. In each round, the information was randomly separated into 80% for the preparation of the ML models, using five-overlay cross-approval, with the remaining 20% being held back for the self-ruling testing. During the cross-approval, the recursive element end way of doing things was completed for the part-related choices inside the cross-approval. That eliminates the highlights most capable of being damaged, while trying to eliminate the conditions and collinearity that may exist in the model. Also, there are many hyperparameters related with every ML model, and the choice on the estimates of hyperparameters is significant. A part of the significant hyperparameters is discussed here.

2.4.3 Neural Network for Treatment Procedure

In their nature and use, neural networks are just like linear inverse models. They contain input and output nodes, use connection weights, bias weights, and cross to learn or train a model. Artificial Neural Networks (ANN) learn tasks by using inductive learning sets of computer instructions to organize huge datasets. A working paper [26] on the use of ANN in decision support systems states that the structure, quality, and amount of data used is critical for the learning process, and that the chosen attributes must be complete, measurable, and independent. Business computer programs or external data sources are usually employed to add to internal data sources.

2.4.3.1 Classification and Prediction Modeling

The classification is defined as the process of finding a model that describes and distinguishes data classes or ideas based on analysis of a set of training data. The models, called classifiers, are categorical class labels and can be used to label the class of objects for which the class label is unknown. Furthermore, the process is described to consist of a learning and a classification step, when a model is used to class labels for a given dataset. Methods include Bayesian classification, SVM, and *K*-Nearest-Neighbor classification. Applications can broadly include detection of illegal stealing, target marketing, performance, manufacturing, and medical.

The available data are divided into two sets for cross-validation: a training set, used to develop a model, and a test set, accustomed to the model's performance. Appropriate data splitting is a way of doing things commonly used in machine learning in order to improve models. Using more training data improves the classification model, whereas using more test data adds to guessing error. Although a 70:30 ratio can usually be used for the relative training and testing sizes, different sampling ways of doing things, ranging from simple to more pre-decided, can be used to split the data, depending on the goals and complex difficulty of the problem.

The classification categorical labels, moving backward is used to obtain missing or unavailable number-based data values. The authors describe Moving Backward analysis as a way of doing things, which is often used for number-based data values and includes identification of distribution, based on available data. An example of number-based data values is when a model is built to a continuous-valued function or ordered value. Such a model usually uses Moving Backward analysis.

ANN can be used in modeling to provide new alternatives to logistic moving backward, the most commonly used method for developing models for results in medicine. Users require less formal training, and the networks can detect complex non-linear relationships and interactions between dependent and independent variables, that cause other things to change. ANN can combine and incorporate literature-based and experimental data to solve problems [27]. Other advantages of ANN, relative to traditional modeling ways of doing things, include fast and straightforward operation due to the compact representation of knowledge, the ability to operate with noisy or missing information, and to generalize to almost the same hidden data, the ability to learn inductively from training data, and the non-linear ability of the process to do very important things when dealing with real-word data.

2.4.3.2 Data Collection

Examining and testing is carried out so that a decision can be made for a problem which happened in two stages. The identified problem was imported to a reference manager and to Covidence, a web-based software, for examining and testing, so that a decision could be made, which was included in something and, when something was kept out or not included, judging requirements were built repetitively *via* agreement. Titles and abstracts were first screened to include articles with keywords related to and/or in clear reference to neural networks.

2.4.4 Prediction of Cancer Susceptibility

From our examination of the literature, a few patterns were noted. As has been commented on already, the use of AI in the prediction of malignant cancers is developing quickly, with the number of papers on the topic expanding by 25% every year. Whereas the use of AI computer programs in disease forecasting and estimations are obviously developing, so too is the use of standard measurably based smart models [28].

When looking at the kinds of expectations or estimates being made, by far the most are related to predicting malignant growth and disease development. Not prevented by part of the issue, a developing number of later indicators are now pointed toward predicting the event of harmful cancer or the risk factors related with developing the disease. When in doubt, paying little attention to the AI used, the sort of forecast being made or the kind of disease being evaluated, the AI ways of doing things seem to improve the quality of expectations.

In surveying how these forecasts were made, the majority examined were dependent on medical-based malignant cancer or segment information (age, weight, smoking, e.g.), either alone or in combination with other sub-atomic biomarkers. Whereas histological information is commonly more open, the doctor who is an expert on diseases with many histopathological symptoms quite often finds it hard to sum up or to move an AI device, prepared on this sort of information, to other medical-based settings. Given the restrictions of using histological judgments in AI, there is a growing power to permitting pattern among the latter to use the more strongly countable highlights, for example, such as clear protein markers and quality changes. Around 47% of studies used this sub-atomic information, either alone or in combination with medicine-based information. Given the most sub-atomic measures, we accept that the results from these examinations should be all the more effectively or powerfully able to move to other medicine-based settings [29].

2.5 CONCLUSION

A wide assortment of exciting and futuristic uses of AI procedures and stages, in the area of medical care, have been discussed in this chapter. Themes ranging from radiology to sensible health tasks, from customized medication to computerized assessment for general health, were described. Known problems of information security and legal structures will continue to be obstructions to the full usage of these frameworks. It may very well be incredibly intricate to make sense of what sort of information can be seen and utilized lawfully by outsider suppliers. Subsequently, an enormous justification exercise of the legal and strategic approaches is required, in addressing those difficulties. As technologists and AI experts, we ought to make progress toward a splendid future where the intensity of AI calculations benefit billions of average citizens to improve their fundamental health and prosperity.

In summary, critical advances have been made in the field of imaging informatics in medicine-based practice, beginning with the limit of examples for whole-slide imaging and stretching out to PC-based understandings. A few advances, such as multispectral imaging, make it conceivable to distinguish and follow changes in countable edges, which may provide new indicative signs that are not clear by human visual evaluation alone. Whereas up until now there have been important obstacles and concerns related to the combination of imaging with the current medicine-based practice, the way forward, with the development of PC power and the increased degree of recognition of its value among doctors, who are experts in diseases, makes it likely that these ways of doing things will turn out to be extremely important for some common medicine-based systems in the not-so-distant future. Meanwhile, the field of imaging informatics has achieved many significant applications in widespread use today.

REFERENCES

1. S.H. El-Sappagh, S. El-Masri, A.M. Riad and M. Elmogy. "Data Mining and Knowledge Discovery: Applications, Techniques, Challenges and Process Models in Healthcare." *International Journal of Engineering Research and Applications (IJERA)*, vol. 3, pp. 900–906, May-June 2013.

2. A.J. Aljaaf, D. Al-Jumeily, A.J. Hussain, P. Fergus, M. Al-Jumaily and K. Abdel-Aziz. "Toward an Optimal Use of Artificial Intelligence Techniques within a Clinical Decision Support System," in *Proc. Ther Science and Information Conference (SAI)*, 2015, pp. 548–554.
3. K.R. Foster, R. Robert Koprowski and J.D. Skufca. "Machine Learning, Medical Diagnosis, and Biomedical Engineering Research Commentary." *BioMedical Engineering OnLine*, vol. 13, p. 94, 2014.
4. D.E. Robbins, V.P. Gurupur and J. Tanik. "Information Architecture ofa Clinical Decision Support System," in *Proceedings of IEEE, Southeastcon*, 2011, pp. 374–378.
5. A.E. Smith, C.D. Nugent and S.I. McClean. "Implementation of Intelligent Decision Support Systems in Health Care." *Journal of Management in Medicine*, vol. 16, pp. 206–218, 2002.
6. D. Foster, C. McGregor and S. El-Masri. "A Survey of Agent-Based Intelligent Decision Support Systems to Support Clinical Management and Research.", in *Proc. International Conference on Autonomous Agents and Multiagent Systems*, 2005.
7. N. Gupta, A. Rawal, V.L. Narasimhan and S. Shiwani. "Accuracy, Sensitivity and Specificity Measurement of Various Classification Techniques on Healthcare Data." *IOSR Journal of Computer Engineering (IOSR-JCE)*, vol. 11, pp. 70–73, May–June 2013.
8. M. Sewak and S. Singh. "In Pursuit of the Best Artificial Neural Network for Predicting the Most Complex Data," in *Proc. International Conference on Communication, Information & Computing Technology(ICCICT)*, 2015, pp. 1–6.
9. A. Dastanpour, S. Ibrahim, R. Mashinchi and A. Selamat. "Comparisonof Genetic Algorithm Optimization on Artificial Neural Network and Support Vector Machine in Intrusion Detection System," in *Proc. IEEE Conference on Open Systems (ICOS)*, 2014, pp. 72–77.
10. A. Al-Rahayfeh and M. Faezipour. "Classifiers Comparison for a NewEye Gaze Direction Classification System," in *Proc. Systems, Applications and Technology Conference (LISAT)*, 2014, pp. 1–6.
11. S.D. Thepade and M.M. Kalbhor. "Novel Data Mining based Image Classification with Bayes, Tree, Rule, Lazy and Function Classifiers using Fractional Row Mean of Cosine, Sine and Walsh Column Transformed Images," in *Proc. International Conference on Communication, Information & Computing Technology (ICCICT)*, 2015, pp. 1–6.
12. I.H. Witten, E. Frank and M.A. Hall. *Data Mining: Practical Machine Learning Tools and Techniques*. Burlington: Morgan Kaufmann, 2011.
13. K. Bache and M. Lichman. "The UCI Machine Learning Repository." http://archive.ics.uci.edu/ml, April, 04, 2013.
14. D.W. Aha. "Cleveland Clinic Foundation Heart Disease DataSet." http://archive.ics.uci.edu/ml/datasets/Heart+Disease, April, 04, 2013 [May, 18, 2015].
15. R.S. Forsyth. "BUPA Medical Research Ltd, Liver Disorders DataSet." https://archive.ics.uci.edu/ml/datasets/Liver+Disorders, April, 04, 2013 [May, 20, 2015].
16. C.V. Hedvat. "Digital Microscopy: Past, Present, and Future." *Archives of Pathology & Laboratory Medicine*, vol. 134 (11), pp. 1666–1670, 2010.
17. J. Pinco, R.A. Goulart, C.N. Otis, J. Garb, L. Pantanowitz. "Impact of Digital Image Manipulation in Cytology." *Archives of Pathology & Laboratory Medicine*, vol. 133 (1), pp. 57–61, 2009.
18. Y. Yagi, J.R. Gilbertson. "The Importance of Optical Optimization in Whole Slide Imaging (WSI) and Digital Pathology Imaging." *Diagnostic Pathology*, 2008.
19. J.D. Martina, C. Simmons, D.M. Jukic. "High-Definition Hematoxylin and Eosin Staining in a Transition to Digital Pathology." *Journal of Pathology Informatics*, vol. 2, p. 45, 2011.

20. M. Thrall, L. Pantanowitz, W. Khalbuss. "Telecytology: Clinical Applications, Current Challenges, and Future Benefits." *Journal of Pathology Informatics*, vol. 2, p. 51, 2011.
21. K. Yamashiro, K. Taira, S. Matsubayashi, et al. "Comparison Between a Traditional Single Still Image and a Multiframe Video Image Along the z-Axis of the Same Microscopic Field of Interest in Cytology: Which Does Contribute to Telecytology?" *Diagnostic Cytopathology*, vol. 37 (10), pp. 727–731, 2009.
22. L.H. Koch, J.N. Lampros, L.K. Delong, et al. "Randomized Comparison of Virtual Microscopy and Traditional Glass Microscopy in Diagnostic Accuracy Among Dermatology and Pathology Residents." *Human Pathology*, vol. 40 (5), pp. 662–667, 2009.
23. P.A. Bautista, Y. Yagi. "Improving the Visualization and Detection of Tissue Folds in Whole Slide Images through Color Enhancement." *Journal of Pathology Informatics*, vol. 1, p. 25, 2010.
24. P.A. Bautista, Y. Yagi. "Detection of Tissue Folds in Whole Slide Images." In *Conference Proceedings: 31st Annual International Conference of the IEEE EMBS*, 2009. pp. 3669–3672.
25. J. Ho, A.V. Parwani, D.M. Jukic, et al. Use of Whole Slide Imaging in Surgical Pathology Quality Assurance: Design and Pilot Validation Studies." *Human Pathology*, vol. 37 (3), pp. 322–331, 2006.
26. D.C. Wilbur, K. Madi, R.B. Colvin, et al. "Whole-Slide Imaging Digital Pathology as a Platform for Teleconsultation." *Archives of Pathology & Laboratory Medicine*, vol. 133 (12), pp. 1949–1953, 2009.
27. P.S. Nielsen, J. Lindebjerg, J. Rasmussen, et al. "Virtual Microscopy: An Evaluation of Its Validity and Diagnostic Performance in Routine Histologic Diagnosis of Skin Tumors." *Human Pathology*, vol. 41 (12), pp. 1770–1776, 2010.
28. F. Wang, J. Kong, L. Cooper, et al. "A Data Model and Database for High-Resolution Pathology Analytical Image Informatics." *Journal of Pathology Informatics*, vol. 2, p. 32, 2011.
29. K. Furuya, T. Maeda, S. Nakanura, T. Kikuchi. "The Role of Computer-Aided 3-Dimensional Analytic Tools and Virtual Microscopy in the Investigation of Radiologic-Pathologic Correlation." *Archives of Pathology & Laboratory Medicine*, vol. 133 (6), pp. 912–915, 2009.

3 Foundation of Machine Learning-Based Data Classification Techniques for Health Care

Bindu Babu, S. Sudha, and L. Godlin Atlas

CONTENTS

3.1 INTRODUCTION

Machine Learning (ML) is the most common technique for predicting the future or classifying knowledge to help people make necessary decisions. ML techniques are based on algorithms that explain the relationship between variables. ML algorithms are trained in instances where they learn from past data and can even evaluate historical data. After continuous training, such an algorithm can recognize patterns to make predictions. ML provides tools, techniques, and services and can aid in various medical fields to overcome predictive, diagnostic, and therapeutic issues. This is used to assess the significance of clinical features and their combinations for diagnosis, e.g., appropriate disease prediction, medical information extraction for testing, medication and aid, and overall patient management. ML is also used in data analysis, such as detecting anomalies in fragmented data, labeling continuous data used in the intensive care unit (ICU) scheduling, and the Smart Alert that results in reliable and effective tracking. Successful adoption of ML strategies can help integrate computer-based technologies into the healthcare environment, establishing possibilities for supporting and improving the practice of medical professionals, and improving quality and service performance. Figure 3.1 illustrates the general workflow of ML algorithms. The four different phases in ML are 1) ML data gathering and preprocessing, 2) building of datasets, 3) training and 4) ML evaluation. Initially, data are gathered from different sources and aggregated into a single dataset, then preprocessed to convert them into a usable dataset. Processed datasets are divided into a training dataset, a validation dataset, and a testing dataset. The training set is then given to the appropriate ML algorithm for the latter to learn features used in classification. After completing the training, the model can be refined using the validation dataset, in which the acceptable level of individual parameters is found. Once the parameters are located, testing is done using the testing dataset.

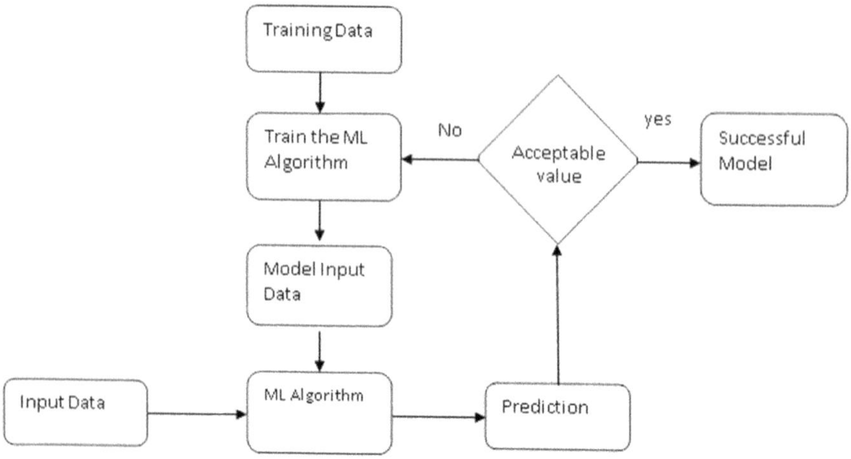

FIGURE 3.1 General workflow of machine learning algorithms.

If the results are acceptable, then the algorithm can be provided for prediction, otherwise training must be continued to improve the accuracy.

3.2 MACHINE LEARNING TECHNIQUES

ML techniques are of three types: supervised learning, unsupervised learning, and reinforcement learning, as shown in Figure 3.2. Supervised learning is used to learn mapping from the input variable 'x' to the output variable 'y.' Such models know the algorithm from the training datasets. It is analogous to a teacher supervising the learning process. Unsupervised learning algorithms identify data based on their densities, structures, comparable fragments, and similar features. In reinforcement learning, the agent experiments with the system, and the system responds to this experimentation with benefits or penalties. The agent, at that point, advances its conduct. The objective is to maximize the rewards and limit the punishments.

3.3 SUPERVISED LEARNING

The supervised learning algorithm uses a training set and known output responses to develop a model to create a reliable prediction for new input data.

Supervised learning involves training a model that relates an individual's characteristics and habits to a specific outcome (e.g., prediction of a disease). Supervised learning models make predictions that will be either discrete (e.g., cancerous/non-cancerous) or continuous. A supervised model that generates discrete categories (sometimes referred to as classes) is referred to as a classification algorithm. Examples include whether the tumor detected is non-cancerous or cancerous, or whether the patient's written remarks reflect positive or negative feelings. In practice,

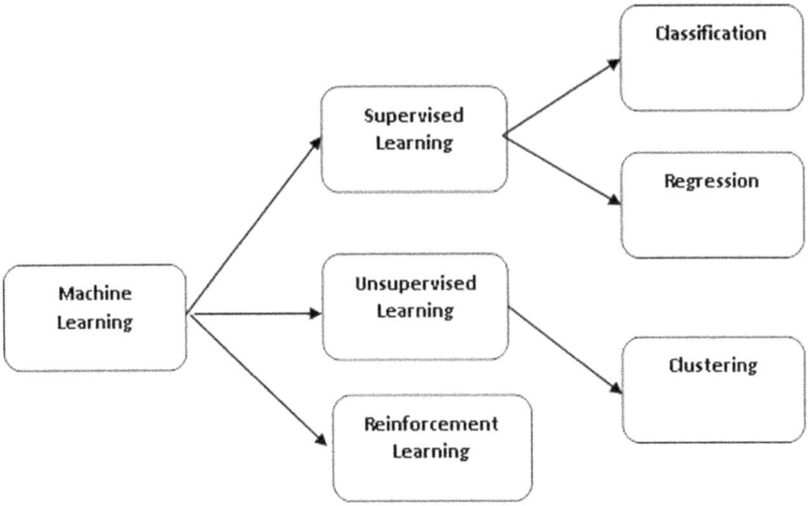

FIGURE 3.2 Machine learning techniques.

classification algorithms return the probability of a class (0 for impossible or 1 for certain). A supervised model that produces continuous value estimations is called a regression algorithm. Using such an algorithm, one can estimate an individual's lifespan or the acceptable dose of chemotherapy. Supervised ML algorithms are usually constructed using a dataset containing a set of training variables and a corresponding output. In contrast with supervised learning, unsupervised learning does not involve the training part.

3.4 CLASSIFICATION

Classification is the method of separating the datasets into various categories or classes by adding a label. It assigns each data point to a specific group based on a particular criterion. Classification is done to carry out predictive analysis on the dataset. Different classification methods are used in various medical fields, such as the estimation of the efficacy of surgical operations, medical tests, medications, and investigation of the relationship between clinical and diagnostic data. Classification algorithms commonly used are Decision Tree, Logistic regression, *K*-Nearest Neighbor (KNN), Support Vector Machine (SVM), Neural Network, and Naïve Bayes.

3.4.1 Decision Tree

The decision tree-based supervised learning method is a graphical representation of all possible alternatives to a decision. This is a predictive model used to graphically arrange the information on future choices, outcomes, and end value. Decision trees identify new data by grouping them, based on feature values. Here, data are categorized according to class, and results are represented as a tree structure. This supervised learning model classifies data by moving through the 'yes' or 'no' structure from the roots to the leaf, considered to be one class. Every node in the decision tree represents a value that can be inferred by the node. The data are categorized, beginning at the root node and sorted, based on their characteristic values, before eventually entering the leaf node.

The below-mentioned steps are used to classify the data using decision trees.

1. Decision tree brings all examples of learning to a root.
2. It separates the teaching samples based on the characteristics chosen.
3. It uses specific mathematical methods to select attributes.
4. Iterative partitioning occurs until instances of training are not in the same type.

Consider an example where a patient wants to have hip replacement surgery carried out [4]. He has two options.

1. Total hip replacement – the whole joint is replaced.
2. Hip resurfacing – only part of the joint is replaced, allowing normal hip function.

The decision tree predicts which type of hip replacement provides a probability of good recovery.

Decision tree building begins with a decision node, where a decision is made, and the lines or branches of the decision tree extending from this node show the options at this point.

It is trailed by a chance node, which characterizes a hazard and what will occur. Additionally, transition probabilities are given, which are the probabilities of moving into the endurance branch or demise branch. At every terminal node, payoffs are defined, which can be healthcare cost or health outcome. Here, it is described as good work and bad work. The payoff for good work is one, and the payoff for bad work and demise is 0.

The better replacement technique can be found out by measuring the expected estimation of the probability for both methods [4], as shown in Figures 3.3 and 3.4.

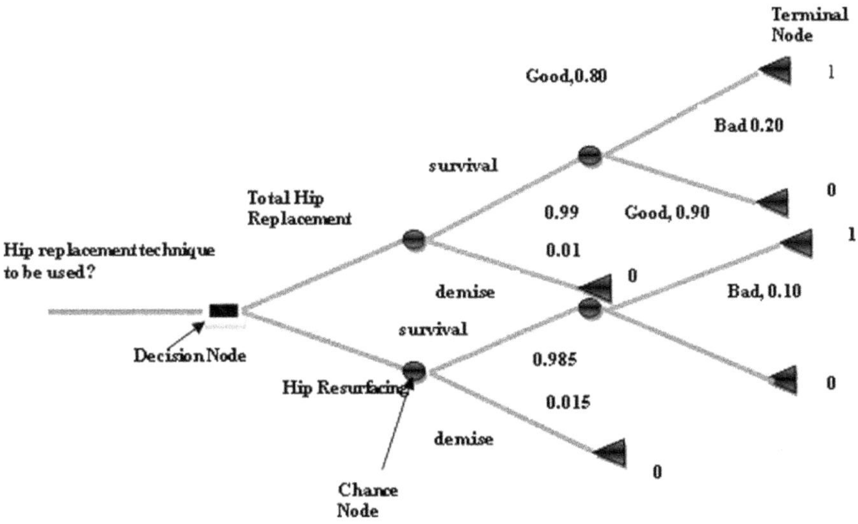

FIGURE 3.3 Selection of hip replacement technique based on decision trees.

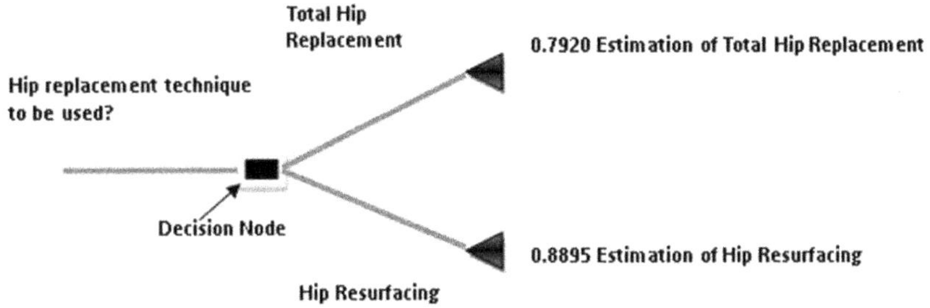

FIGURE 3.4 Likelihood of hip replacement techniques.

First, consider the total hip replacement branch of the tree:

$$\text{Probability of payoff for good work} = \Pr(\text{survival}) \times \Pr(\text{good work}) \times \text{Payoff}$$

$$= 0.99 \times 0.80 \times 1 = 0.7920$$

$$= 0.99 \times 0.20 \times 0.00 = 0.00$$

$$\text{Probability of payoff for demise} = \Pr(\text{demise}) \times \text{Payoff}$$

$$= 0.01 \times 0.00 = 0.00$$

Next consider the hip resurfacing:

$$\text{Probability of payoff for good work} = \Pr(\text{survival}) \times \Pr(\text{good work}) \times \text{Payoff}$$

$$= 0.985 \times 0.90 \times 1 = 0.8895$$

$$\text{Probability of payoff for bad function} = \Pr(\text{survival}) \times \Pr(\text{bad work}) \times \text{Payoff}$$

$$= 0.985 \times 0.10 \times 0.00 = 0.00$$

$$\text{Probability of payoff for demise} = \Pr(\text{demise}) \times \text{Payoff}$$

$$= 0.015 \times 0.00 = 0.00$$

After referring to the expected value from each surgical option, we can see that hip resurfacing has a higher probability of a good recovery.

3.4.2 Random Forest

Random Forest is a series or group of decision trees. Consider a situation where we aim to distinguish people into two categories, such as statin respondents and statin non-respondents. Figure 3.5 shows the classification, using Random Forest. The Random Forest cycle starts with a group of training datasets, consisting of known statin responders and non-responders, each identified by a set of characteristics, such as age, sex, smoking and diabetes status. A group of decision trees is constructed, where each tree uses these prognostic features to classify as either statin responders or non-responders. At each node in each tree, one feature is selected that can achieve this distinction. As we use a single feature in the node, the next nodes are used to achieve the perfect separation. The resulting stochasticity allows each tree to vote independently on the final prediction and acts as a regularization means. Even though an individual decision tree is unlikely to be accurate on its own, the final voting, based on hundreds of trees, will give an accurate result.

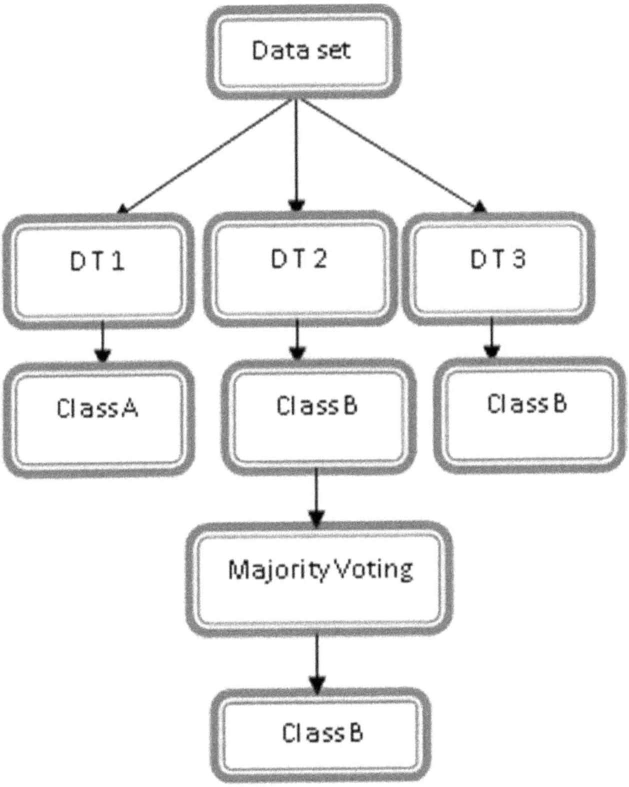

FIGURE 3.5 Classification using Random Forests.

Features of Random Forest

- The most precise learning algorithm
- Works well for classification as well as regression problems
- Works effectively on massive datasets
- No input training is required
- Performs tacit selection of features
- Can be quickly grown in parallel
- Error correction strategies available for unbalanced datasets

3.4.3 KNN Algorithm

K-Nearest Neighbor is a basic algorithm that records every accessible case and classifies new information based on a similarity measure. *K* signifies the number of neighbors closest to the voting class of the most recent information or the test

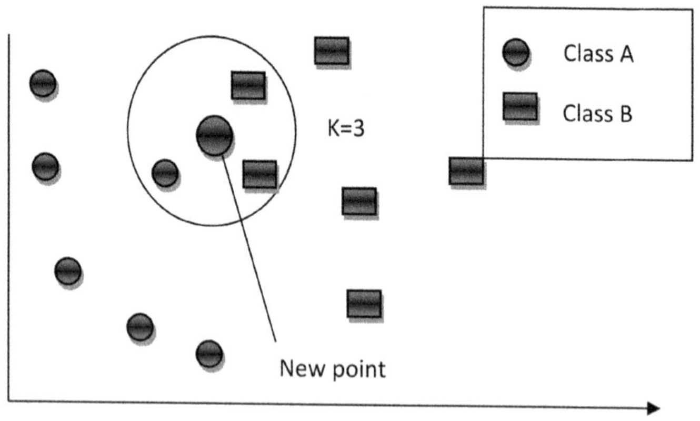

FIGURE 3.6 Prediction using the *K*- Nearest Neighbor (KNN) algorithm.

data. If *K*=1, the test data will be given a similar imprint as the training set's nearest sample.

Let the circles denote Class A, and the squares represent Class B. The function is to decide if the new dot belongs to Class A or Class B. Figure 3.6 shows how prediction is carried out, using the KNN algorithm. When *K*=3, select the three dots nearest to the new dot. Here, the minimum distance between the samples needs to be measured. The minimum distance can be calculated using the Euclidean distance, the Hamming distance, or the Manhattan distance. If counted, we can see from Figure 3.6 that there are two squares and one circle, which implies that the new dot belongs to Class B.

The Euclidean distance is the simplest method to measure the minimum distance between the samples. The Euclidean distance is the square root of the sum of the squares of the difference between the new point and the current one.

The separation between two points $x = (x_1, x_2)$ and $y = (y_1, y_2)$ can be obtained, using the Pythagorean hypothesis:

$$D(x,y) = \sqrt{(x_1 - y_1)^2 + (x_2 - y_2)^2} \tag{3.1}$$

This method can be used for both classification and regression. A typical weighting scheme consists of giving each neighbor a weighting factor of 1/d, where d is the distance to the neighbor so that the closest neighbor contributes more to the average than does the farthest one [1].

The hyperparameter *K*, which represents the number of neighbors, must be selected at the time of model creation. *K* can be used as a variable to monitor the prediction model. The optimal value of *K* for most datasets ranges from 3 to 10.

Steps to evaluate the KNN algorithm:

1. Determine *K*
2. Measure the distance between the test sample and the training samples

3. Calculate the closest neighbors, based on the minimum *K*–th distance
4. Note down the list of the nearest neighbors
5. The mode of the most immediate neighbor group collected is chosen as the predicted value

Studies have shown that the KNN classifier can predict heart diseases with an accuracy of almost 90% [8].

3.4.4 Naïve Bayes

Naïve Bayes is a classification technique that uses Bayes' theorem, assuming individuality among the predictors. Naïve Bayes is a simple yet essential and important algorithm for predictive analysis. Naïve Bayes Classifier assumes that the features in a class are unique, regardless of whether they rely upon one another, with each of these properties autonomously contributing to the likelihood. The naïve-based model is simple to construct and is particularly useful for huge datasets.

Bayes' theorem depicts the likelihood of an event being dependent on earlier information on the conditions that they may be related to the event. It is a way to work out the conditional probability. Conditional probability is the probability that an event will occur, provided that it has a relationship to one or more other events. In mathematical terms, Bayes' theorem states that

$$P(X/Y) = \frac{P(X)\,P(Y/X)}{P(Y)} \tag{3.2}$$

where A and B are the events. P(X) and P(Y) are the prior probabilities of X and *Y* without regard to each other. P(X/Y) is called the posterior probability, which is the probability of observing event X, given that Y is true. P(Y/X) is called the likelihood, which is the probability of observing event Y, given that X is true.

Let user A want to find whether he is suffering from malaria, given that the prior probability of it is as P(X=malaria) = (fatigue, fever, shivering, sweating, muscle pain)/(overall symptoms). Now a person is exhibiting fever and muscle pain, so P(Y=fever and muscle pain), then, given that the probability of having malaria (from the training dataset), what are the chances that this user A is suffering from malaria:

$$P(Y \mid X) = \frac{P(X \mid Y) {*} P(Y)}{P(X)} \; P(X) \text{ should not be } 0 \,(\text{zero}) \tag{3.3}$$

3.4.5 Neural Networks

One of the best-known ML algorithms utilized today to classify and diagnose medical information is Artificial Neural Networks (ANNs). The invention of neural networks took place in the 1970s. ANN is a particular category of ML algorithms that are built like human brains. It works similarly to the neurons in our nervous system, which gain information from past data. Also, ANN learns data and give responses in the form of predictions or classifications.

The human brain comprises an enormous number of interconnected nerve cells called neurons. A neuron consists of a cell nucleus, separate branches called dendrites, and one long fiber called the axon. The dendrites receive message signals from different neurons. The signal then moves to the soma, which is the main cell body. The signal exits the soma and moves to the neurotransmitter down the axon and then to the next nerve cell. By simultaneously using many neurons, the brain can process data and perform its function very rapidly. The working of the brain inspired the ANN.

The neural network has three significant layers, namely the input layers, hidden layers, and output layers, as shown in Figure 3.7. The input layer is the primary layer of an ANN, that obtains the input information as various texts, numbers, audio files, picture pixels, and so forth. Hidden layers are the center layers of the ANN. There can be single or multiple hidden layers. The hidden layers perform various sorts of mathematical computations on the input data and recognize the associated patterns. The outcome acquired is shown in the output layer.

There are many parameters in the neural network, such as weights, biases, learning rate, etc., that influence the model's performance. Figure 3.8 shows the architecture of neural networks. Every node has a weight associated with it. A transfer function is used to measure the weighted sum of the inputs and the bias.

$$\sum_{i=1}^{n} w_i x_i = w_1 x_1 + w_2 x_2 + \ldots.. w_n x_n \tag{3.4}$$

After the weighted sum is calculated, the activation function provides an output '1' if the weighted sum is greater than the pre-threshold value; otherwise, the outcome will be '0'. Sigmoid, RELU, Softmax, and tanh are commonly used activation functions.

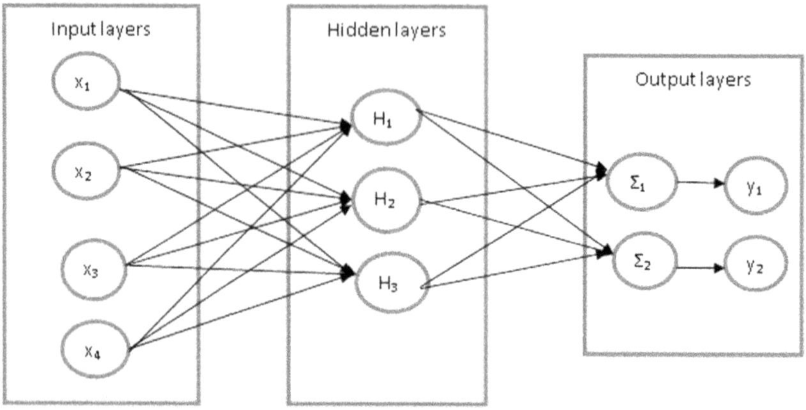

FIGURE 3.7 Feed-forward Artificial Neural Network.

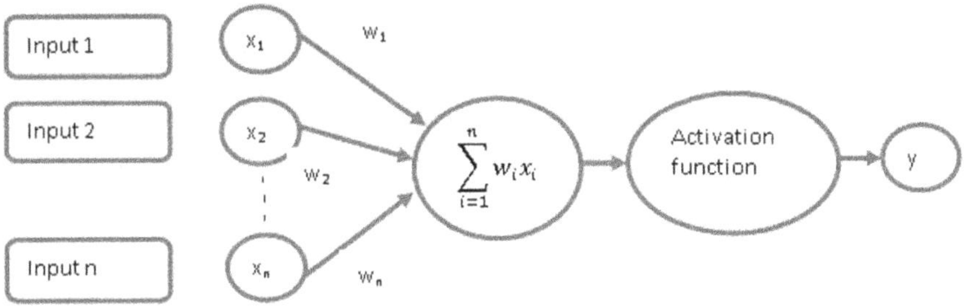

FIGURE 3.8 Architecture of neural networks.

The types of neural networks include

Feed-forward networks: Feed-forward network-based ANN allow signals to travel in only one direction, i.e., from input to output, and there is no feedback from any layer. These types of neural networks are extensively used in pattern recognition.

Feedback networks: ANN with feedback networks can have signals traveling in both directions by introducing feedback loops in the network. Feedback networks are dynamic; their state changes continuously until they reach an equilibrium point. They stay at the equilibrium point until the information changes, when another equilibrium should be found.

3.4.5.1 Back Propagation in ANN

For a neural network to be trained, examples of input–output mappings are given to the network. Finally, after the neural networks finish the preparation, the neural network can be checked without these mappings. The neural network estimates performance, and the prediction quality is tested using different error functions. Finally, depending on the test, the model alters the neural network's weights to improve the system.

The possible uses of neural networks in the healthcare sector include diagnostic systems, biochemical analysis, image processing, and drug production. ANNs help provide health predictions that doctors and surgeons cannot address on their own. ANNs function at moments where data can be processed but without understanding which aspects of the data are critically relevant. ANNs can identify complex connections that probably would not be obvious initially, but will lead to better public health predictions.

3.4.6 Support Vector Machines (SVM)

SVM are supervised learning models that evaluate information and characterize patterns for classification and regression analysis. This model is mainly used for labeling purposes. SVM classifiers distinguish the two classes by using a decision boundary called the hyperplane. With regard to a set of training data where each data point is categorized into one of the two classes, the SVM training algorithm fabricates a model that assigns new data points to one class or the other. The plotting is performed in n-dimensional space. The value of each element is also the value of the unique coordinate. So, the perfect hyperplane is to be sought that distinguishes between the two groups. An SVM model portrays samples as space points, represented in such a way as to show the examples in different groups by a marginal space that is as wide as possible. New samples are then mapped to the same space and projected to belong to a class, depending on which side of the gap they fall [28]. In addition to doing linear classifications, SVMs can effectively execute a nonlinear classification, using kernel techniques, i.e., transforming their inputs into high-dimensional feature spaces indirectly.

Consider N vectors $\{x_1, x_2,\ldots, x_n\}$, with each vector x_j having m features $\{a_{j1}, a_{j2},\ldots, a_{jm}\}$ and belonging to one of two classes Y_1 or Y_2, if it is possible to determine a hyperplane in the feature space that can distinguish between the examples from class Y_1 and class Y_2. In Figure 3.9, linearly separable data (circles and squares) are shown in a two-dimensional feature space. The discrete line represents the separating hyperplane, which is a straight line in two dimensions. Suppose the given data are not linearly separable in the m-dimensional feature space. In that case, it is possible to map it into a higher-dimensional space, making the data linearly separable, using a kernel function. The SVM kernel is a function that converts low-dimensional input space into higher-dimensional space, i.e., transforms non-separable into separable. Because of the linearly separable data, there will be a gap between the two examples of the given classes. This gap is referred to as the margin. It would be advantageous if the margin between the two classes were as wide as possible.

Consider 'm' feature vectors that are chosen to be the training set for the given 'n' vectors. The real target of the SVM is to discover a hyperplane that groups all

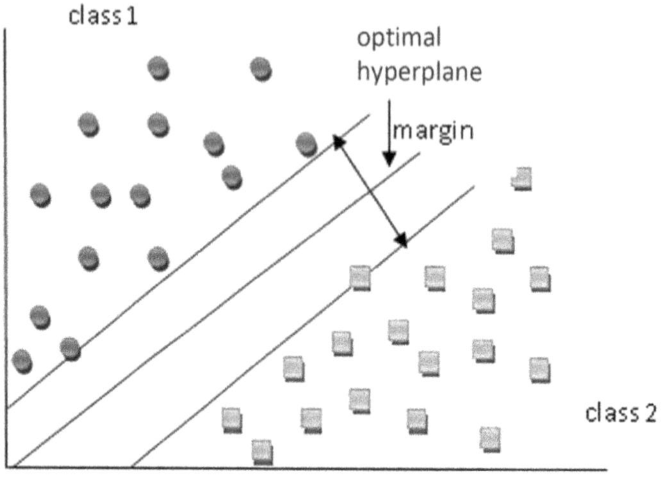

FIGURE 3.9 Linearly separable data of circles and squares in a two-dimensional feature space.

training vectors into two classes. There might be a lot of separating hyperplanes between the two classes. Data points closest to the hyperplane are referred to as support vectors. Hyperplanes will be defined based on these instances. Hyperplanes should be selected so that the distance margin between the two classes is maximal; otherwise, there is a chance that new cases will be incorrectly classified. A good margin arises when all the support vectors have the same distance within the hyperplane's maximum margin.

The SVM is a linear discriminant function of the form:

$$S(x) = \mathbf{w}^T\mathbf{x} + b \tag{3.5}$$

x is a feature vector, w is an m-dimensional weighting vector, and the scalar b is a bias. The weighting vector is orthogonal to the hyperplane and controls its direction, whereas the bias controls its position.

$$\text{If } S(\vec{x}) = \mathbf{w}^T\vec{x} + b > 0; \vec{x} \text{ is an instance in } Y_{1.} \tag{3.6}$$

$$\text{If } S(\vec{x}) = \mathbf{w}^T\vec{x} + b < 0; \vec{x} \text{ is an instance in } Y_{2.} \tag{3.7}$$

At the SVM training stage, the weight and the bias are adjusted so that all the examples of Y_1 lie on one side of the hyperplane, and all the examples of Y_2 lie on the other side of the hyperplane.

3.4.6.1 Advantages of SVM Classifiers

The SVM classifiers offer high accuracy and function admirably within a high-dimensional space. The SVM classifiers fundamentally utilize only a subset of training points, so the outcome is that they use substantially less memory.

3.4.6.2 Disadvantages of SVM Classifiers

They have a high preparation time; in this manner, they are not appropriate for enormous datasets. Another disadvantage is that SVM classifiers do not function properly when the classes are non-separable.

3.4.6.3 Applications of SVM

1. Subtle pattern recognition in complex datasets
2. Handwriting recognition
3. Fraudulent credit card recognition
4. Speaker identification and facial recognition
5. Cancer genomics – discovery of new biomarkers, new drug targets, understanding of cancer driver genes [5].

3.5 REGRESSION

Regression analysis is a type of predictive modeling technique that analyzes the correlation between a dependent variable (target) and an independent variable (predictor). It involves calculating the equation of the line-of-best-fit over a scatter of data points that most closely matches the overall shape of the data. It shows the changes of the dependent or y variable to the changes in the explanatory (independent) variable, x. The three critical applications of the regression analysis are the estimation of the power of the predictors, the forecasting of results, and the forecasting of patterns in which the regression analysis forecasts trends and future values. There are different types of regression algorithms: linear regression, logistic regression, polynomial regression, etc. The logistic regression algorithm is the one most used to classify data in the healthcare sector. In this chapter, logistic regression will be discussed in detail.

3.5.1 Logistic Regression

A logistic regression algorithm is used to allocate observations to a discrete set of classes. It can be used to solve the classification problems, like classifying whether the email is spam or not, whether an online transaction is fraudulent or not, or whether a tumor is cancerous or non-cancerous. The logistic regression output should be discrete/categorical, for example, 0 or 1, yes or no, true or false, or high or low. Logistic regression transforms its predictions using the logistic sigmoid function to a probabilistic value between 0 and 1, as shown in Figure 3.10. In logistic regression, a threshold is set. If the data point is below the threshold point, the output is 0; otherwise, 1.

The sigmoid function is expressed as

$$\sigma(z) = \frac{1}{1-e^{-z}}; \text{where z} = \sum w_i x_i + \text{bias} \tag{3.8}$$

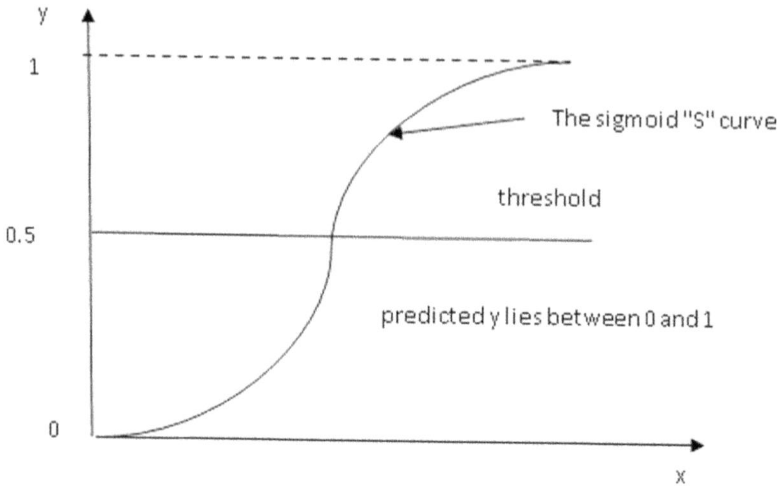

FIGURE 3.10 Logistic regression curve.

Logistic regression can be used to help to assess the disease. Consider a patient heading to the hospital to get a regular check-up. The doctor will conduct different tests on the patient to determine whether the patient is sick. The doctor will register parameters, such as sugar level, blood pressure, heart rate, age, medical history, etc. Eventually, the doctor assesses the patient's records and arrives at a conclusion as to whether the patient is ill or not.

3.6 UNSUPERVISED LEARNING

Unsupervised learning is a type of ML technique in which no one monitors or guides the model. Instead, the model works on its own to determine the information. The unsupervised learning technique deals with unlabeled data, i.e., there will be only input data and no corresponding output variables. It helps discover unknown patterns in the data, and even helps to find the features to be used for categorization.

3.7 CLUSTERING

Clustering is an essential principle in unsupervised learning when it comes to healthcare data. The main aim is to find a structure or pattern in the set of unmarked data. It divides the dataset into different classes, such that the data in the same cluster have the highest correlations among one another [26]. Clustering is used in numerous applications, such as pattern recognition, image processing, and consumer purchase pattern analysis. Nowadays, clustering methods are commonly used in the healthcare industry for fast diagnosis and disease prediction.

Consider the situation where a ML model predicts a patient's chances of having a heart attack. In most cases, the output (the label of whether or not a person has had a heart attack) is associated with inputs (parameters like age, weight, blood pressure, heart rate, previous history, etc.) A supervised classification model, like SVM, can be used to learn the patterns associated with it. Then, if a new patient entry occurs, the trained classifier can predict whether the patient has a heart problem or not. But if the reliable outputs are not given, the categorization must be done from the input values alone. In such a condition, clustering techniques can help to carry out the classification.

Some investigators have merged clustering algorithms with classifying models to achieve successful classification and diagnosis, to overcome the difference between supervised and unsupervised learning [6]; for example, the Wisconsin Diagnostic Breast Cancer Dataset from the University of California has been used to model breast cancer diagnostic algorithms [9]. They used a fusion method of *K*-means and SVMs. The *K*-means algorithm was used to reveal the underlying patterns of malignant and benign tumors, which could be distinguished by using the *K*-means algorithm. At the same time, the detection of breast cancer was achieved using the SVM classifier. These hybrid approaches have demonstrated a high degree of precision. Consequently, clustering methods must also be shown to enhance the classification of results. Among the unsupervised learning algorithms, clustering by *K*-means is the one most widely used due to its simple concept.

3.7.1 *K*-Means Clustering

The function of *K*-means clustering is to define a group of *K* clusters, so that each data point is assigned to the nearest cluster. Consider an example to achieve a better understanding of the topic. We have several black dots in the figure below, and we aim to allocate them to three clusters. The *K*-means follow the primary four phases listed below.

Phase 1: Initiate Cluster Centers [26]

Figure 3.11 shows the data points before clustering. The three data points, O_1, O_2, and O_3, are selected randomly and shown as circle, star, and triangle, respectively, to denote cluster centers as shown in Figure 3.12.

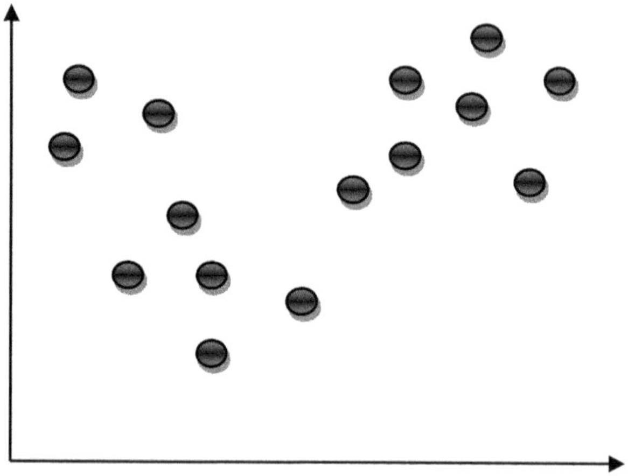

FIGURE 3.11 Data points before clustering.

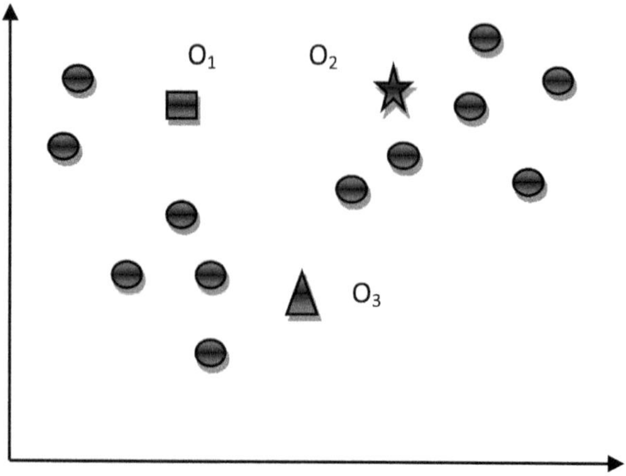

FIGURE 3.12 Assigning cluster centers.

Phase 2: Allocate Findings to the Nearest Cluster Centers

Once cluster centers are located, each data point is allocated to a cluster, based on the minimum distance from the point to the cluster center. Calculate the distance from the black point A to O_1, O_2, and O_3, respectively, as shown in Figure 3.13.

When data points l_1, l_2, and l_3 are compared, it is calculated that l_1 is the shortest, so point A is assigned to the square cluster and labelled with the square color. The same protocol is followed at all stages. The first stage of cluster formation is shown in Figure 3.14.

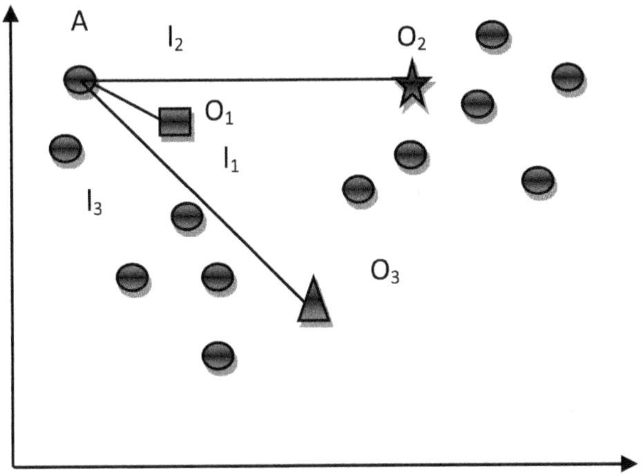

FIGURE 3.13 Distance calculation from cluster centers.

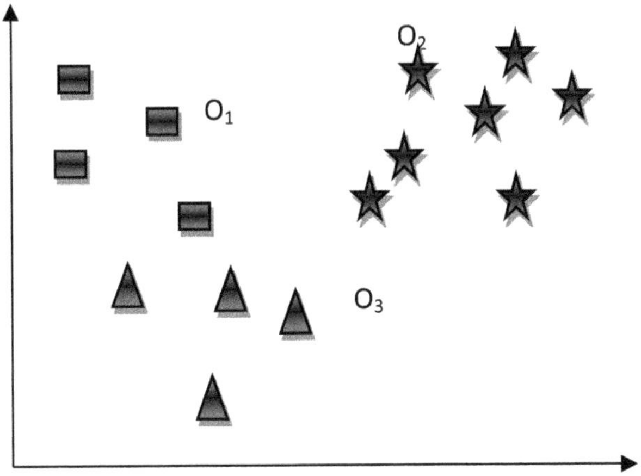

FIGURE 3.14 First stage of cluster formation.

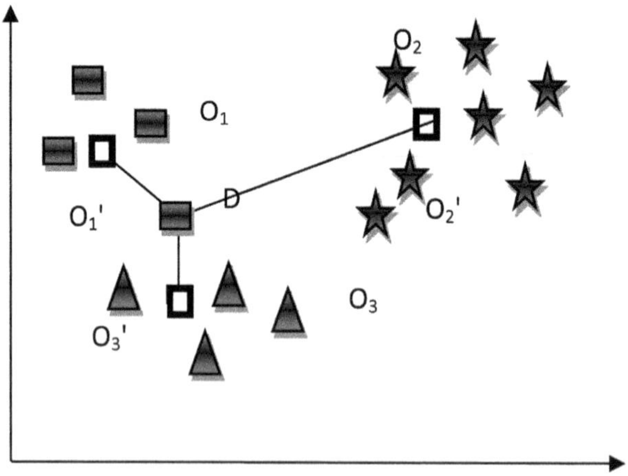

FIGURE 3.15 Assigning new cluster centers.

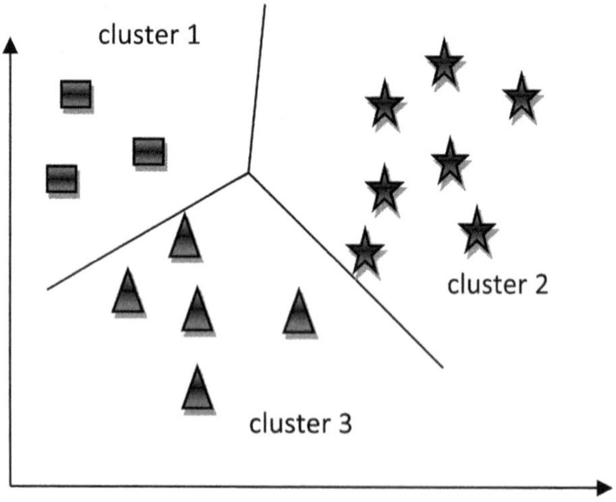

FIGURE 3.16 Final clustering.

Phase 3: Reexamine the Cluster Centers by Taking the Mean of the Assigned Observations

All data points are linked to clusters, based on the cluster center to which they are closest. The next phase is to upgrade the cluster centers. The center mass of the square clusters is calculated by summing up all square data points and dividing by the total number of square points, which is four here. The resulting center mass is O_1', the new center for square clusters. Similarly, new centers O_2' and O_3' for star and triangle clusters, respectively, which can be found in the same way, as shown in Figure 3.15.

Phase 4: Repeat Phases 2 and 3 Until Convergence

The last section in the *K*-means clustering is to repeat phase 1 and phase 2. For example, O_1', O_2', O_3' are known as new cluster centers, and point D is similar to triangle clusters. Point D can, therefore, be marked as a point in the triangle cluster. This cycle is replicated until convergence has been achieved. Final clustering is shown in Figure 3.16.

3.8 APPLICATIONS OF MACHINE LEARNING IN HEALTHCARE

3.8.1 Patterns of Imaging Analytics

Healthcare organizations worldwide are progressively engaged with developing image analysis and pathology with the aid of ML software and algorithms. ML systems help radiologists to recognize subtle differences in scans, permitting them to distinguish, identify, and analyze early-stage health problems. Another such weighty advance is Google's AI calculation to classify malignant tumors in mammograms.

3.8.2 Personalized Treatment

The penetration rate of Electronic Health Records in health care has increased since 2012, inferring increased and more noteworthy access to individual patient health records. Through analyzing this personal clinical information from individual patients with ML software and algorithms, healthcare providers can classify and determine health problems.

3.8.3 Discovery and Manufacturing of Drugs

ML algorithms are also successful in drug development, especially at the early level, from the initial screening process of pre-drug compounds to the predicted success rate, based on biological factors.

3.8.4 Identifying Diseases and Diagnosis

Due to the new techniques like ML and deep learning, physicians can now even detect certain diseases that were previously beyond detection, such as tumors or cancers in the early stages of genetic diseases.

3.8.5 Robotic Surgery

As a result of robotic surgery, doctors can operate effectively and with precision even under the most challenging conditions. Robotic surgery is also commonly used in hair transplantation, as this process requires more delicate detailing. Robotics, powered by AI and ML algorithms, improve the accuracy of surgical instruments but also integrate real-time surgical measurements, data from previous surgical experience, and pre-op medical records into the surgical procedure.

3.8.6 Clinical Trial Research

ML systems provide a wide variety of ways to enhance clinical trial science. By applying smart predictive analytics to clinical trial candidates, medical practitioners can evaluate a more extensive dataset, that would reduce the costs and time needed to perform medical trials.

3.8.7 Predicting Epidemic Outbreaks

Healthcare organizations apply ML algorithms to track and predict future outbreaks of an epidemic that can impact various parts of the globe [2, 3]. These tools can predict epidemic outbreaks by gathering data from satellites, real-time social media alerts, and other critical web-based information.

3.8.8 Improved Radiotherapy

There are many discrete variables involved in medical image processing that can be stimulated at any random moment. ML calculations gain from a few specific information tests. They can also analyze and characterize the desired variables. For instance, ML is utilized in medical image analysis to identify artifacts, like lesions, into different categories – normal, abnormal, lesion-free, benign, malignant, etc.

3.8.9 Maintaining Healthcare Records

It is understood that routinely reviewing and preserving health records and patient medical history is an extensive and expensive process. Developments in the ML calculations help tackle this issue by diminishing the time, effort, and assets associated with the record-keeping process. Data classification approaches utilizing VMs (vector machines) and ML-based optical character recognition (OCR) techniques, such as the Google Cloud Vision API, help arrange and order healthcare information.

3.9 CONCLUSION

The development of ML in the healthcare field is impressive, from optimizing the delivery structure of healthcare services, reducing costs, and handling patient data more efficiently to developing new medication and drug treatments, remote monitoring, and so on. The selection of the correct ML algorithm for an application is essential. Through continued advances in data science and ML, the healthcare industry can now harness innovative technologies to provide quality treatment.

REFERENCES

1. Wedyan, Mohammad, Alessandro Crippa, and Adel Al-Jumaily. "A novel virtual sample generation method to overcome the small sample size problem in computer aided medical diagnosing." *Algorithms* 12.8 (2019): 160.
2. Toussie, Danielle, et al. "Clinical and chest radiography features determine patient outcomes in young and middle age adults with COVID-19." *Radiology* 197–205 (2020): 201754.
3. Puntmann, Valentina O., et al. "Outcomes of cardiovascular magnetic resonance imaging in patients recently recovered from coronavirus disease 2019 (COVID-19)." *JAMA Cardiology* 5 (2020): 1–9.
4. Grampurohit, Sneha, et al. "Disease prediction using machine learning algorithms." In *2020 International Conference for Emerging Technology (INCET)* Belgaum, IEEE, (2020), 1–7.
5. Gonzalez-Hernandez, Jose-Luis, et al. "Technology, application and potential of dynamic breast thermography for the detection of breast cancer." *International Journal of Heat and Mass Transfer* 131 (2019): 558–573.
6. Visvikis, Dimitris, et al. "Artificial intelligence, machine (deep) learning and radio (geno) mics: definitions and nuclear medicine imaging applications." *European Journal of Nuclear Medicine and Molecular Imaging* 2630–2637 (2019): 1–8.
7. Bashar, Abul. "Survey on evolving deep learning neural network architectures." *Journal of Artificial Intelligence* 1.02 (2019): 73–82.
8. Shouman, Mai, Tim Turner, and Rob Stocker. "Applying k-nearest neighbour in diagnosing heart disease patients." *International Journal of Information and Education Technology* 2.3 (2012): 220–223.
9. Alshehri, Abdulelah S., Rafiqul Gani, and Fengqi You. "Deep Learning and Knowledge-Based Methods for Computer Aided Molecular Design--Toward a Unified Approach: State-of-the-Art and Future Directions." arXiv preprint arXiv:2005.08968 (2020).
10. Mikołajczyk, Agnieszka, and Michał Grochowski. "Style transfer-based image synthesis as an efficient regularization technique in deep learning." In *2019 24th International Conference on Methods and Models in Automation and Robotics (MMAR).* IEEE (2019).
11. Sahoo, Abhaya Kumar, Chittaranjan Pradhan, and Himansu Das. "Performance evaluation of different machine learning methods and deep-learning based convolutional neural network for health decision making." In *Nature Inspired Computing for Data Science.* Springer, Cham (2020), 201–212.
12. Nebbia, Giacomo, et al. "Deep learning of sub-regional breast parenchyma in mammograms for localized breast cancer risk prediction." In *Journal of Medical Imaging 2019: Computer-Aided Diagnosis.* Vol. 10950. International Society for Optics and Photonics (2019).
13. Shahnaz, Ayesha, Usman Qamar, and Ayesha Khalid. "Using blockchain for electronic health records." *IEEE Access* 7 (2019): 147782–147795.
14. Shakeel, P. Mohamed, et al. "Neural network based brain tumor detection using wireless infrared imaging sensor." *IEEE Access* 7 (2019): 5577–5588.
15. Tsohou, Aggeliki, et al. "Privacy, security, legal and technology acceptance requirements for a GDPR compliance platform." In *Computers & Security.* Sokratis, Katsikas (Ed.) Springer, Cham (2019), pp. 204–223.
16. Harrer, Stefan, et al. "Artificial intelligence for clinical trial design." *Trends in Pharmacological Sciences* 40 (2019): 577–591.

17. Padala, Suresh, and Veerababu Padala. "Fusion of CT and MRI scanned medical images using image processing." *Journal of Computer Technology & Applications* 3.3 (2019): 17–20.
18. Reynaert, N. "PET and MRI based RT treatment planning: Handling uncertainties." *Cancer/Radiothérapie* (2019): 173–192.
19. Stankovic, Zoran, Bradley D. Allen, Julio Garcia, Kelly B. Jarvis, Michael Markl. "4D flow imaging with MRI". *Cardiovascular Diagnosis and Therapy* 4.2 (2014): 173–192. doi: 10.3978/j.issn.2223-3652.2014.01.02.
20. Kim, Yejin, et al. "Federated tensor factorization for computational phenotyping." In *Proceedings of the 23rd ACM SIGKDD International Conference on Knowledge Discovery and Data Mining* (2017).
21. Mohammed, Mohssen, Muhammad Badruddin Khan, and Eihab Bashier Mohammed Bashier. *Machine Learning Algorithms And Applications.* Taylor & Francis Group, LLC (2017).
22. Bohra, Himdeep, Amol Arora, Piyush Gaikwad, Rushabh Bhand, and Manisha R. Patil. *International Journal Of Advanced Research In Computer And Communication Engineering* ISO 3297:2007 Certified 6.4(2017).
23. Maithili, A., R. Vasantha Kumari, and S. Rajamanickam. Neural Networks Towards Medical. *International Journal of Modern Engineering Research (IJMER)* 1.1: 57–64.
24. Wang S, and Cai Y. "Identification of the functional alteration signatures across different cancer types with support vector machine and feature analysis." *Biochim Biophys Acta Molecular Basis Disease* 1864.6 Pt B (2018): 2218–2227.
25. Huang, Shujun, Nianguang Cai, Pedro Penzuti Pacheco, Shavira Narandes, Yang Wang, and Wayne Xu. "Applications of support vector machine (SVM) learning in cancer genomics." *Cancer Genomics Proteomics* 15.1 (2018): 41–51.
26. Ogbuabor, Godwin and F. N. Ugwoke. "Clustering algorithm for a healthcare dataset using silhouette score value." *International Journal of Computer Science & Information Technology (IJCSIT)* 10.2 (2018): 27–37.
27. Han, J., M. Kamber, and J. Pei. Cluster Analysis-10: Basic Concepts and Methods (2012).
28. Zheng, B., S. W. Yoon, and S. S. Lam. "Breast cancer diagnosis based on feature extraction using a hybrid of K-means and support vector machine algorithms." *Expert Systems With Applications*, 41.4 (2014): 1476–1482.

4 Deep Learning for Computer-Aided Medical Diagnosis

Sreenarayanan N. M., Deepa C. M., Arjun K. P., and Neethu Narayanan

CONTENTS

4.1 INTRODUCTION

This is the era where developments in deep learning and the accessibility of large annotated clinical image datasets are prompting dramatic changes in obtaining information from medical images. From this point of view, it is necessary to analyze how machine-aided diagnosis of clinical/medical images has contributed to our ability to diagnose and how these advances in technology are going to impact on our society.

Deep networks naturally integrate features and classifiers in a multilayer fashion, with the feature levels being improved by the number of stacked layers. The depth of the network is a major factor in analyzing various results. Since the successful implementation of ImageNet Classification, based on convolutional neural networks, researchers have started being actively involved in the rapidly developing field of deep learning. The major goal of this chapter is to provide an overview of (1) the changes which have been made by the introduction of deep learning to health care, (2) the major deep learning models, and (3) the applications of deep learning models. Machine learning with feature inputs was a major component of computer-aided medical imaging before the introduction of deep learning; the main difference between machine learning before and after the inception of deep learning is the systematic learning of data imagery directly without segmentation. There are two major models in the class of machine learning in medical imaging, massive training artificial neural network (MTANN) and convolutional neural network (CNN), which have similarities as well as a number of differences. Appropriate use of statistical methods in different learning models will facilitate the easy diagnosis of diseases.

Progress in deep learning has opened up a new era that can naturally identify diagnostic highlights from an dataset. Numerous clinical activities have been expedited by the use of deep learning over the past decade. The emergence of large-scale clinical datasets will highlight the value of deep learning in obtaining clinical findings.

4.2 COMPUTER-AIDED MEDICAL DIAGNOSIS

With the increasing adoption of neuro-imaging scanners in medical clinics and organizations, the work of radiologists is expanding. The manual examination of clinical images suffers from errors and intra-radiologist variation in interpretation. Moreover, the element of subjectivity will negatively impact the output of manual image interpretation. Computer-aided medical diagnosis [1] (CAMD) is a methodology to help clinical specialists in the interpretation of clinical images, which may originate from X-ray, computed tomography (CT), ultrasound, magnetic resonance imaging (MRI), positron emission tomography (PET), single-photon emission computed tomography (SPECT), or thermography. CAMD can assist radiologists by achieving interpretation of a clinical image in a fraction of a second.

Ongoing advances in the field of deep learning opens up another system that can naturally identify highlights from the enormous dataset of images. Clinical projects in the area of deep learning, developed over the past decade, such as Human Cerebrum Venture, Brain Initiative, Blue Brain Project, and so on, have provided enormous information in this area.

4.2.1 Radiography

Radiography is one of the main imaging strategies used in present-day medicine. It utilizes wide light-emission beams to create the image. These images help in the evaluation of the appearance/non-appearance of disease, damage, or foreign objects [2].

Two types of radiographic images are utilized in clinical imaging.

1. **Fluoroscopy:** Produces a continuous stream of internal images of the human body parts, yet requires the consistent contribution from lower dose X-ray beams. Accordingly, basically used in picture-controlling strategies, where static and continuous feedback are requested by the strategy.
2. **Projectional Radiography:** X-ray beams are widely used to decide or detect the sort and degree of the fracture and to detect changes in the human bones or lungs. They are used to visualize internal zones around the stomach and digestive tract, and can help in diagnosing major ulcers and specific types of colon cancer.

4.2.2 Magnetic Resonance Imaging (MRI)

The MRI scanner utilizes powerful magnets, discharging radio-recurrent pulses at the specific resonant frequency of hydrogen particles to energize and excite hydrogen cores of normal water atoms in human tissue [3]. The MRI scan uses neither X-rays nor ionizing radiation. The MRI scan is generally utilized in clinics and is seen as achieving superior decisions than a computed tomography (CT) scan because MRI helps in clinical analysis without subjecting the body to direct radiation [4]. MRI checks take a longer time and are more accurate. In addition, individuals with clinical implants or non-removable metal items inside the body cannot go safely through MRI checks.

4.2.3 Ultrasound

Ultrasound utilizes high-intensity ultrasonic waves in the 2 to 18 MHz limit range, which are reflected by tissue to varying degrees to create a type of three-dimensional (3D) image. It is principally associated with fetal monitoring in a pregnant woman. Ultrasound is similarly used for the imaging of the stomach, breast, heart, ligaments, muscles, veins, and arteries. It gives less anatomical detail than is achieved with CT or MRI scans. The main advantage is that ultrasound imaging helps with continuous examination, with the capacity to monitor moving structures without transmitting any ionizing radiation. Extremely safe to use, ultrasound can be performed immediately, without any side-effects and the cost is relatively modest.

4.2.4 Thermography

Thermographic cameras detect long-infra-red (IR) radiation produced by the body, generating heat-sensitive images dependent on the radiation received. Digital infrared thermal imaging (DITI) is the type of thermography that is utilized to analyze breast cancer [5]. The cameras used for thermography will recognize radiation in the long-infra-red range of the electromagnetic spectrum and create images of that radiation, known as thermograms. The measure of various adiation increments increases with rising temperatures. Accordingly, thermography helps to identify variation in temperature. It is capable of imaging moving items continuously. The significant weaknesses of this clinical imaging method are that:

- Thermographic cameras are very costly.
- Images of the items which exhibiting temperature variation, will probably not result in any precise warm images of itself.

4.2.5 Nuclear Medicine Imaging (NMI)

This sort of clinical imaging is widely carried out by taking radio dense chemicals internally. External gamma detectors capture the radiation, generating images from the radio chemicals [6]. This is the inverse of X-rays where the radiation passes through the body from the outside, with, in this NMI scenario, the gamma rays are discharged from inside the human body. In any case, the radiation dose is low enough to pose no risk.

4.3 DEEP LEARNING IN HEALTH CARE

Deep learning (DL), machine learning (ML), artificial intelligence (AI), and semantic computing have been the major methods in data analysis in the clinical industry. Each one of these innovations is associated with providing support to clinical staff. Because of fruitful trial results and assorted applications, DL can possibly change the fate of the healthcare industry. AI procedures are utilized for detecting malignancies and monitoring treatment [16]. Systemic DL techniques for computer vision achieves even more exact clinical imaging and interpretation.

4.3.1 Deep Learning Neural Network

DL provides the healthcare industry with the ability to analyze information from images at a remarkable speed without compromising on accuracy. It is not only ML,

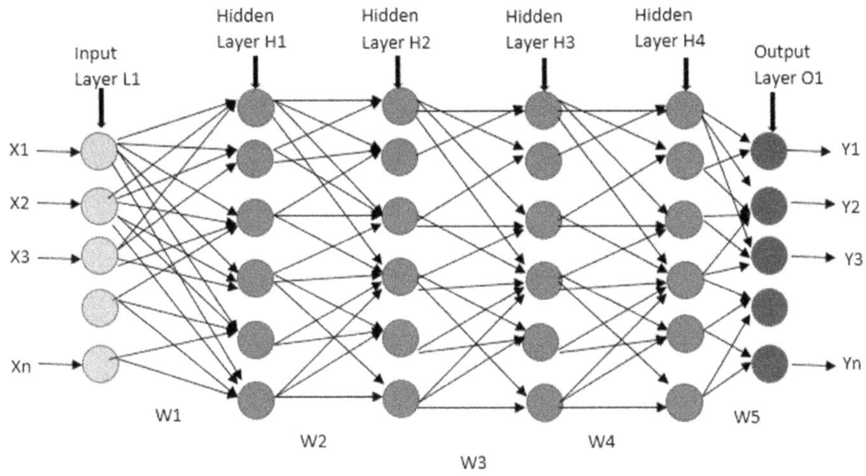

FIGURE 4.1 An illustration of a deep learning neural network.

nor is it AI, but it is a rich mixture of both, utilizing a different layered and structured algorithmic engineering to filter vital information at a surprisingly high speed [7]. The major key to the focus of DL is in the name – "learning".

In DL models, information is sifted through a course of different layers, with each progressive layer utilizing the information from the previous one to inform its outcomes. DL models can become increasingly more precise as they measure more information, gaining from past outcomes to refine their capacity to identify relationships and associations. Similar to the manner in which electrical signals traverse living cells, each network layer might carry out a particular segment of a task, and the information may navigate the layers on various occasions to refine and advance a definitive output (Figure 4.1). These "hidden" layers will play out the various interpretations that transform crude contributions into the important target output.

4.3.2 Imaging Analytics and Diagnostics

One type of CNN may be appropriate to investigating images, such as those generated by X-rays or MRI scans. CNNs are designed to handle images, as mentioned by specialists at Stanford University, allowing the networks to work more efficiently and to handle larger numbers of images. Therefore, some CNNs are approaching – or, in some cases, outperforming – the accuracy of human diagnosticians, while recognizing and diagnosing significant highlights in analytic imaging.

Convolutional Neural Networks are developed to investigate dermatology images, recognized melanoma with 10% greater accuracy than human clinicians. Furthermore, when clinicians were provided with major foundation data on patients, for example, sex, age, and the body site of the potential melanoma, the CNN still outperformed the dermatologists by almost 7%. In addition to being exceptionally precise, decisions from deep learning systems are also made very rapidly.

A group from the Nvidia Corporation graphics technology company, the Massachusetts General Hospita, the Mayo Clinic, and the Brigham's and Women's Hospital Center for Medical Data Science has built up a strategy for utilizing generative adversarial networks (GANs), another kind of DL, which can make extremely useful clinical images. The images use designs developed from genuine outputs to make CT or MRI scan images [18]. The data allows specialists to access enormous volumes of vital information without concerns around patient privacy or consent. These reproduced images are exact to the point that they can help to train future deep learning models to analyze clinical discoveries.

4.3.3 Natural Language Processing in Health Care

Progress in Natural Language Processing (NLP) [8], a part of computerized intelligence that permits computers to comprehend spoken or composed comments, is encouraging medical service associations to become involved in that field.

For example, voice-assistance operatives and dialogue acknowledgment stages, NLP is improving clinical experiences by higher access to patient data, reducing record expenses and delays, and improving the nature of health records.

Elements driving the use of NLP in medical services are as follows:

1. Management of the surge in volume of clinical data
2. Support of the value-based care and population health management
3. Improvement of the patient–provider interactions with electronic health records (EHR)
4. Ability to address and track the need for higher quality of health care

NLP in the medical services industry can help to improve the precision and completeness of EHRs by changing the free-form content into a standardized informational format. This could likewise make documentation simpler by permitting care suppliers to access direct notes, as NLP transforms it into archived information.

4.3.4 Drug Discovery and Precision Medicine

Drug discovery and precision medicine are also in the plans for DL engineers. These targets require handling of enormous volumes of clinical, genomic, and population level information with the objective of recognizing previously obscure relationships between the genes, drugs, and physical conditions. DL is a system for drug partners and scientists hoping to feature new examples in these relatively unexplored informational datasets, particularly with respect to the development of precision medication analysis.

In the world of genotype-specific medication, surprising disclosures are commonplace, making it an exciting testing ground for demonstrating creative ways in which to deal with targeted consideration. This strategy will ideally reveal new information into, say, why and how certain tumors develop in specific patients. DL advances will accelerate the essential task of analyzing such information.

Another report, carried out by specialists from the University of Massachusetts, discovered that systemic DL could likewise recognize adverse drug events (ADEs) with greater precision than could customary models. The system combines DL with natural language handling to sift through unstructured EHR information, to identify problematic relationships between drug type, recurrence, and dose. The outcomes could then be utilized to check the safety of novel treatments or to see whether new drugs are being endorsed in real clinical climates.

4.3.5 Clinical Decision Support Systems

Computer-aided DL may soon be a helpful companion in an inpatient setting, where it can make suppliers aware of changes in high-hazard conditions, such as sepsis and respiratory failure [9]. Analysts from the MIT Computer Science and Artificial Intelligence Laboratory (CSAIL) have developed a system called ICU Intervene [9], which offers clinicians an itemized reasoning for its suggestions, helping to encourage trust.

4.4 DEEP LEARNING VS CAMD

In CAMD, AI techniques are used to analyze information from past case tests from a patient population to build up a model, relating the data with certain disease results [9]. The model created should be capable of predicting the result of another case study when information from another case is introduced into the framework as information. With the assistance of an appropriate prepared framework, the CAMD prediction might be utilized as supporting data in a clinical expert's decision-making cycle. The methodology of utilizing DL innovation to analyze patient information to help decision making is relevant to any patient treatment measure. Moreover, the imaging information assumes a significant function in every one of these cases, so that image examination is a principal part of CAMD.

4.4.1 CAMD for Neurodegenerative Diseases

Neurodegenerative disorders are types of illnesses that involve the breakdown/death of neurons in the brain. Neurodegenerative diseases result in the devastating effects of miscommunication between brain tissue cells. Parkinson's disease, Huntington's disease, and dementia are some examples of neurodegenerative diseases. Neurofunctional imaging is of great importance in clinical assessment. The methodologies and techniques involved in CAMD help to achieve early detection of neurodegenerative diseases. Combined efforts of CAMD and deep neural network architectures yield accurate results for early detection of the disease, ensuring good quality of life for such patients.

Preprocessing of the neurofunctional images is an important step in CAMD systems as it ensures that all the images are comparable. The individual images are mapped from their own subject space to a reference space for the comparison. This process is known as spatial normalization or regularization, incorporating supervised and unsupervised learning methods to identify the symptoms.

FIGURE 4.2 Illustration of overfitting.

4.4.2 Deep Learning and Regularization Techniques

When dealing with huge volumes of clinical records, one of the most general problems that data science experts face is avoidance of overfitting. As it is related to sensitivity, data accuracy is very important. Sometimes, our model will work with trained data, but it fails to predict new data because of anomalies [10]. DL regularization techniques can be used to avoid overfitting of the data, as shown in Figure 4.2. As shown in this figure, we can observe that the complexity of the model increases when going towards the right side, such that the system training error decreases but the system testing error does not.

4.4.2.1 Multi-Task Learning

Multi-task learning is an AI method whereby numerous connected undertakings can be played out at the same time. The framework will link the figures with regard to how to play out the two tasks at the same time with the end goal being that each assignment helps in learning of the other one. This is an approach to emulate human insight, i.e., how people carry out various tasks at the same time. The significant steps in performing multi-task learning are as follows:

1. Creating the dataset
2. Creating the network architecture
3. Defining the multi-task loss function
4. Training

   With respect to point 3, loss function is characterized as the sum of the squared differences between the real value and the anticipated value. Loss function for the linear regression with four information factors will be:

$$L(x,y) = \sum_{i=1}^{n} \left(y_i - f(x_i)\right)^2$$

$$f(x_i) = h_\theta x = \theta_0 + \theta_1 x_1 + \theta_2 x_2^2 + \theta_3 x_3^3 + \theta_4 x_4^4$$

Regularization works dependently on a presumption that lesser load will produce a less- difficult model and assist with abstaining from overfitting.

L1 and L2 are the two normal types of regularization techniques. General cost capacity can be updated by adding another factor, known as the regularization term.

$$\text{Cost function} = \text{Loss}(\text{say, binary cross entropy}) + \text{Regularization term}$$

$$L(x,y) = \sum_{i-1}^{n} \left(y_i - h_\theta(x_i)\right)^2$$

$$\text{where } h_\theta x_i = \theta_0 + \theta_1 x_1 + \theta_2 x_2^2 + \theta_3 x_3^3 + \theta_4 x_4^4$$

$$L(x,y) = \sum_{i-1}^{n} \left(y_i - h_\theta(x_i)\right)^2 + \lambda \sum_{i-1}^{n} \theta_i^2$$

The penalty term or regularization boundary, λ, decides the amount by which to penalize the weightings. At the point when the estimation of λ is 0/zero, then the regularization term becomes zero.

$$L(x,y) = \sum_{i-1}^{n} \left(y_i - h_\theta(x_i)\right)^2 + \lambda \sum_{i-1}^{n} \left|\theta_i^2\right|$$

L1 regularization highlights feature selection. It tends to be accomplished by appointing inconsequential informational highlights with a weighting of zero and helpful highlights with a weighting other than zero. In L2, the regularization term will be the total sums of squares of all component weightings, demonstrated as follows:

$$L(x,y) = \sum_{i-1}^{n} \left(y_i - h_\theta(x_i)\right)^2 + \lambda \sum_{i-1}^{n} \theta_i^2$$

4.4.2.2 Convolutional Neural Network

CNN is a particular neural network widely employed in the field of computer vision. This system plays out a numerical activity called a "convolution". With regards to a CNN, a "convolution" is a dimensional linear operation, which can deal with considerably more information than a conventional neural network [11], given that the method was intended for two-dimensional information and the duplication is performed between an array of information and a two-dimensional array of weightings, called a filter or a kernel. Figure 4.3 summarizes the technique.

The filter is more modest than the information dataset and the kind of duplication applied between a filter-sized fix of the information and the channel is a dot product. A dot product element is the component-wise multiplication between the filter-sized

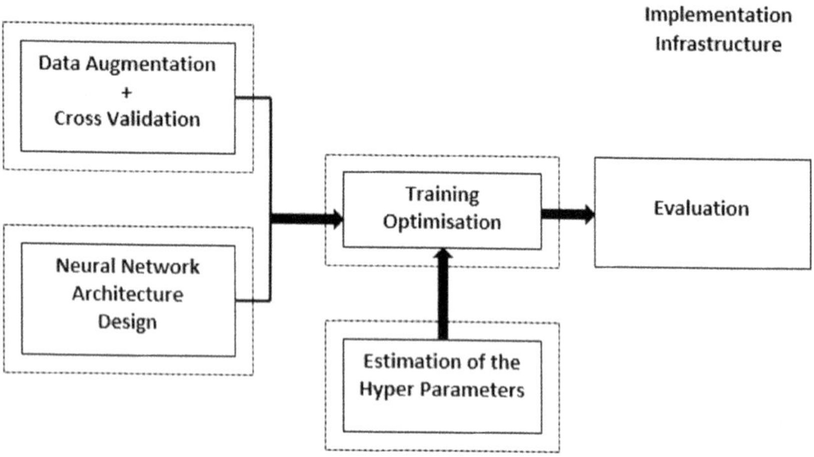

FIGURE 4.3 Procedure of evaluation of a medical image.

fix of the dataset and the filter, which is then added, continually reducing it to a single value. Since it generates a single value, the activity is frequently alluded to as the "scalar item ".

Employing a filter more modest than the dataset is deliberate as it permits a similar filter to be multiplied by the information cluster at various points on the input.

The fully connected feature of the convolutional networks model will make it prone to overfitting of the data. By adding a loss function, regularization methods reduce the effect of overfitting, by increasing the accuracy.

4.4.2.3 Transfer Learning

A neural network is trained on an information dataset. The neural network picks up information from the dataset, which is represented as "weightings" or "loads" of the network. The weightings can be extracted and moved to some other neural network. This will avoid the weightings of preparing the other neural network, as we "transfer" the learned highlights. Transfer learning is a strategy by which to beat isolated learning and to use the information obtained from one task to facilitate other related undertakings.

DL models are examples of inductive learning, as illustrated in Figure 4.4. The fundamental goal for inductive learning calculations is to derive a plan from a collection of training models.

In instances of classification, the trained model works out how to do the planning between input features and the class names. the algorithm works with a series of expectations identified with the spread of the information. These series of expectations are known as inductive bias. This inductive strategy can be described by numerous components, such as the theory space, which limits the search measure through the speculation space. In this way, these inclinations can affect how and what is determined by the model on the given area.

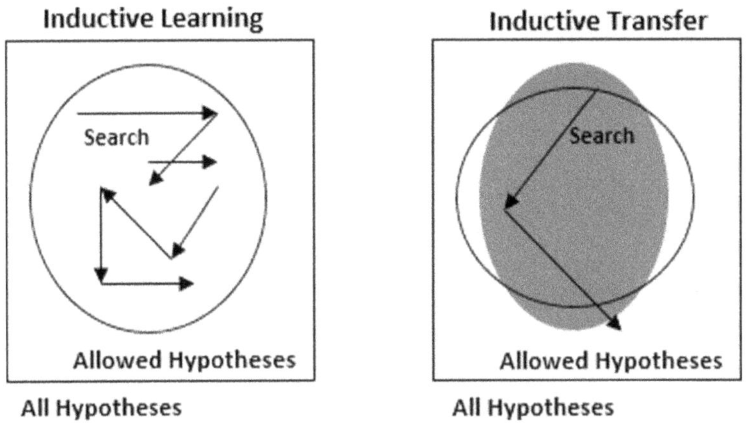

FIGURE 4.4 Inductive learning.

The inductive transfer approach uses the inductive predispositions of the source to help achieve the objective task, by changing the inductive inclination of the objective assignment by limiting the model space, narrowing down the theory space, or making adaptations to the search cycle itself with the assistance of information from the source task. This cycle is shown in Figure 4.4.

4.4.3 CAMD and Big Medical Data

The main challenge in medical imaging is the volume of data generated and the selection of data with which to achieve a good model. Image segmentation is one of the important steps in analyzing MRI (Magnetic Resonance Imaging). Manual segmentation of images is a difficult and tedious task as we are dealing with enormous volumes of data. To clinically diagnose a disease accurately, the amount of data matters. Training the model with a smaller volume of data leads to inaccurate predictions. Computer-aided medical diagnosis and other learning support systems will help to achieve the automatic segmentation of clinical images. Networks like GAN achieved great success in terms of detection, segmentation, and image generation. The DL structure can be made to analyze the images, alongside any decision-support system which will generate an effective predictive model.

The model of treatment is changed, based on the data. Electronic health records (EHRs) are one of the most widespread applications, in which there will a digitalized record of the history of each patient. Medical imaging consists of a variety of image acquisition methodologies. Analytic pipelines can be used for the proper integration of huge volumes of data.

4.4.4 Deep Learning for Cancer Location

There has been an enormous increase in the number of cancer patients seen by clinicians each day. Irregular development of tissue cells of any human body part,

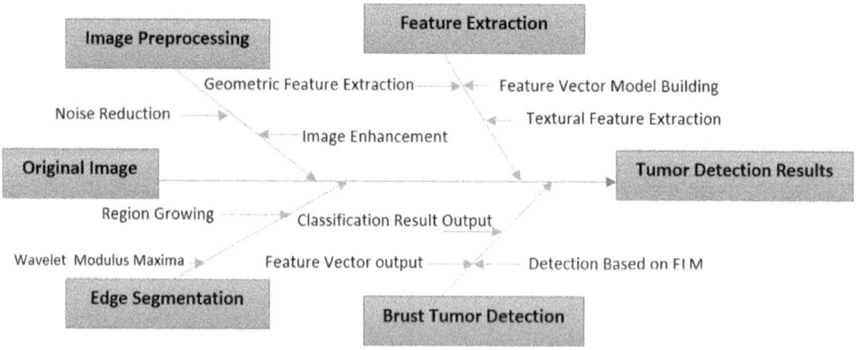

FIGURE 4.5 The framework overview of breast tumor detection.

making up an undifferentiated mass of tissue, is called a "tumor" or "neoplasm". For the most part, normal cells in the human body go through a pattern of division, aging, dying, and replacement by new cells [12]. The occurrence of a tumor causes this typical cycle to be disrupted. There are two kinds of tumor: benign (non-harmful) and malignant (dangerous). A benign tumor is not particularly risky and remains in one part of the body, without spreading to other parts. On the other hand, cells from malignant tumors can metastasize, spreading to other sites of the body, where they generate additional numbers of malignant tumors. The rapid spread of the malignant tumor makes both treatment and prognosis difficult. Figure 4.5 summarizes the detection of breast tumors.

4.5 DL APPLICATIONS IN HEALTH CARE

DL in medical care has begun to leave its imprint. Google has invested a lot of time examining how DL models can be utilized to make predictions concerning hospitalized patients, supporting clinicians in analyzing and integrating patient information and results. The future of medical care has never been more exciting. AI and ML present an opportunity to create arrangements that cater for quite specific requirements inside the industry, whereas DL in medical services can turn out to be extremely powerful in supporting clinicians and improving patient outcomes.

4.5.1 Electronic Health Records

EHR frameworks effectively store and secure patient information, such as demographic data, clinical history data records, and laboratory test results. EHR frameworks improve the speed at which the correct diagnosis can be arrived at and reduce the time it takes to arrive at a diagnosis, by means of the use of DL algorithms [13]. These learning algorithms use information stored in EHR frameworks to identify patterns and high-risk factors, and to make determinations based on the examples they recognize. Scientists can use information from EHR frameworks to develop DL static models, which will anticipate the probability of occurrence of certain human health-related results, such as the likelihood that a patient will develop a particular disease.

1. A static prediction discloses the probability of a function dependent on an informational dataset which scientists feed into the framework and the code embeddings from the "International Statistical Classification of Diseases and Related Health Problems" (ICD). Scientists developed a DL model which classified patients from EHR information, explicitly their clinical history and the rate of their emergency clinic visits.
2. An expectation dependent on an input of information sources uses data from the EHR framework to achieve a forecast. It is conceivable to make a valid prediction with either each information source or with the entire informational index. Scientists created the Doctor AI model, that uses artificial neural networks (ANN) to specifically predict when a future medical clinic visit will happen, and the reason behind the visit. They base this expectation on the dataset, including ICD codes assembled from a patient's past clinic visits and the time elapsed since the patient's most-recent visit.

4.5.2 Drug Discovery

DL in medical care helps in the identification of drugs and their development. This innovation will dissect the clinical history of a patient and afterward select the most appropriate treatment. Additionally, this innovation picks up experience from earlier assessments and tests.

4.5.3 Medical Imaging

Clinical imaging techniques, such as MRI checks, CT examinations, and electrocardiography (ECG), are used to analyze serious diseases, such as heart disease, cancer, and brain tumors. In this way, DL encourages specialists to analyze the disease in more detail, providing the patients with the best therapy.

4.5.3.1 Image Analysis to Detect Tumors

The essential task for recognizing tumors is preprocessing, to improve the nature of the MR images and to make them appropriate for additional handling by machine vision systems. Preprocessing assists with improving the explicit boundaries of MR images, such as signal-to-noise ratio, visual appearance [14], elimination of the unnecessary parts of the background, smoothing of the internal parts of the images, and protection of the edges. Different strategies can be used for different type of tumors. For identifying brain tumors, skull stripping can be carried out for the examination of the tumor from the MR images. The process of skull stripping will remove non-brain tissues from the image. The image can be segmented further for locating objects and boundaries.

Image segmentation will assign labels to pixels, such that pixels having the same labels share certain visual characteristics. As a next step, thresholding can be done to convert gray-scale images into binary images, containing all foreground information. Features can be extracted further to produce the output image.

4.5.3.2 Detecting Cancerous Cells

Feature extraction plays a vital role in the identification of tumor cells. Figure 4.6 illustrates a tumor-detecting mechanism. Features can be extracted with the help of statistical parameters like mean, contrast, entropy, and energy, as shown in Figure 4.6. Extracted features can be fed in to support vector machine (SVM) classifiers to achieve better decision making.

4.5.3.3 Detecting Osteoarthritis from an MRI Scan

MRI scans are an integral tool for diagnosing joint inflammation, being more precise than customary X-rays and ultrasound imaging. MRI is more costly than some other imaging tests, so some centers may utilize it only in clinical preliminaries or for assessing specific conditions, for example, spondyloarthritis.

Arthritis is a condition that causes pain and irritation in the joints. Characterizing joint pain early can assist individuals with receiving successful treatment sooner, to moderate the progress of the illness. Osteoarthritis occurs where the tissue that

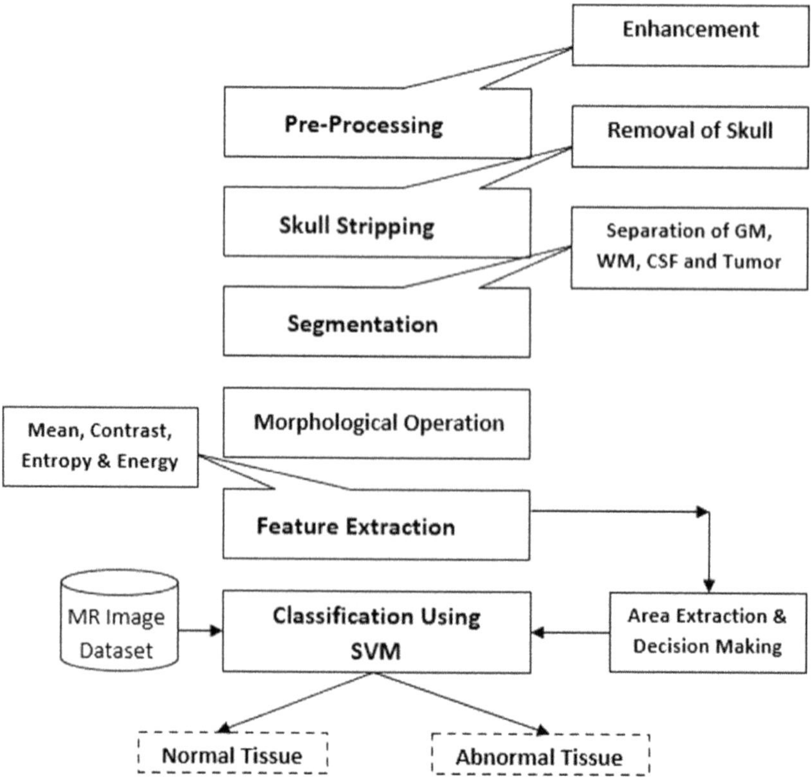

FIGURE 4.6 Procedure for detecting tumor tissues.

covers the bones in joints separates. For the most part, this happens with age or because of an earlier joint physical issue, making the joints rub against one another. An MRI can show bone edema (fluid developing in the bone marrow, which may cause swelling), and inflammation of delicate tissues, as when degenerated cartilage or bone pieces are held up in the joint. HSS employs a convention of explicit MRI pulse groupings to recognize early proof of ligament degeneration [14]. When evidence of ligament wear is recognized at an early stage, treatment can start to prevent or defer progression. This can defer or eliminate the requirement for surgery.

4.5.4 Genome

The DL procedure is used to understand the inherited component underlying the condition. DL has a promising future in genomics, and subsequently in the insurance industry. CellScope at UC Berkeley utilizes DL, enabling parents to screen the strength of their youngsters progressively through the use of a smart gadget. As a consequence, DL will limit the need for regular visits to the specialist. DL in medical care can provide specialists and patients with an array of applications, which will help specialists to improve clinical therapies. Figure 4.7 summarizes the healthcare analytics.

Clinical experts can examine the entire genome of an organism with the use of genomics studies. Comprehensive perspectives on the components (e.g., genes, proteins, mRNAs, metabolites) making up a cell is achieved by omics innovation.

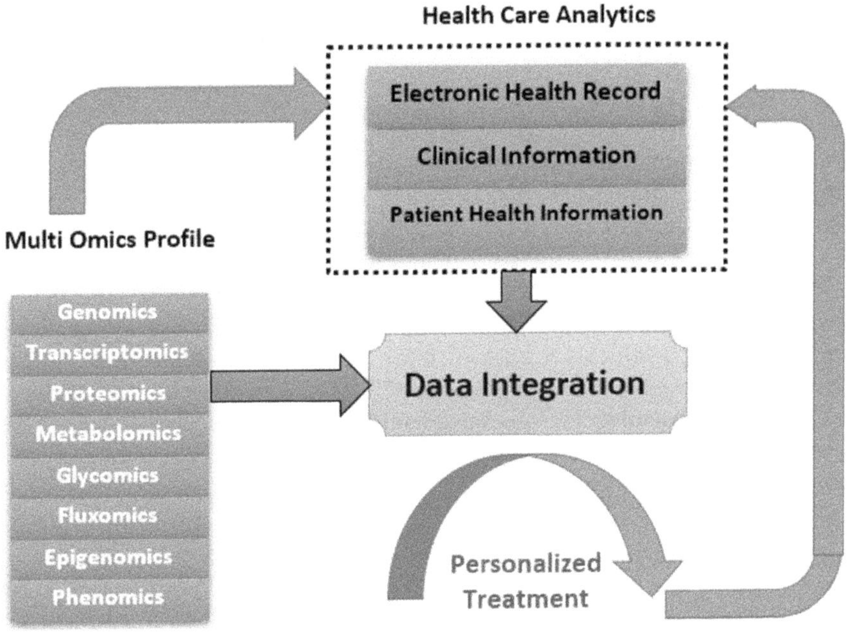

FIGURE 4.7 Healthcare analytics.

The essential point occurs at the universal detection of genes (genomics), metabolites (metabolomics), proteins (proteomics) and mRNA (transcriptomics) in a particular organ in a non-targeted way.

4.5.5 Automatic Treatment

A multi-institutional group in Shanghai has built up a computerized therapy-arranging framework for intensity-modulated radiation treatment, dependent on 3D section expectation and area circulation-based streamlining. They showed that the DL strategies guarantee the production of highly individualized radiotherapy.

For predicting the appropriate dose matrix for a given patient, they built up a network-based structure that can be prepared to associate voxel geometry with voxel area. The information on the remaining network was obtained from the individual trans-axial CT sections of the anatomy and the yield was the dose matrix for each section.

After the training and testing cycles of the DL model, the creators of the framework looked at the anticipated area conveyances and the DVH (Dose–Volume Histogram), with the outcomes determined by the treatment-arranging framework. The outcomes showed that the DL strategy can predict the successful treatments.

4.6 MAJOR CHALLENGES

The major DL challenges facing medical diagnosis are listed in the following subsections.

4.6.1 Limited Dataset

The first and the most significant essential point for the use of DL is the availability of a massive training dataset, as the appropriate quality and assessment of the DL classifier depends strongly on the quality and size of the information dataset. The restricted accessibility of the clinical imaging information is one of the major difficulties facing the achievement of DL innovation.

Improvement of the massive training dataset is itself a difficult and tedious assignment which requires a lot of time from clinical specialists. The availability of more, qualified specialists is expected to make sizeable quality information datasets available, particularly for uncommon illnesses. Also, a reasonable-sized dataset is essential for DL calculations, to achieve familiarity with the underlying characteristics. In medical care, the majority of the accessible datasets are uneven and not in same class.

4.6.2 Privacy and Legal Issues

The sharing of clinical data with the public is a seriously complex and problematic issue, depending on different datasets. Information privacy is as sociological as it is a technical issue, and should be attended to from the two points of view. HIPAA

(Health Insurance Portability and Accountability Act, 1996, US Federal) gives lawful rights to the patients to ensure that their clinical records, individual, and other human health-related data are not given to clinics, health plans, specialists, and other medical services suppliers [15].

With the growth in medical care information, protection of patient data is a major challenge for information science specialists, in light of the fact that disposal of the center's individual data makes the planning of the information very complex; at the same time, a master hacker could gain access to vast quantities of medical information belonging to individual patients.

Differential privacy approaches can be attempted, which limit the information available to individual organizations on a need-to-know basis. Sharing of sensitive information with restricted accessibility is a real test. A great number of restrictions are needed. Restricted access to information, will reduce the risk of such important data. In addition, the information is continually expanding, increasing the risk to information security.

4.6.3 Process Standardization

1. **Information Standards:** The standardization of information is of great importance for learning in any area, particularly for medical services. The purpose for this is that any variation in the information increments from equipment to equipment will lose consistency, causing variation in the information captured. Medical services require aggregations of information from various sources to achieve improved DL and greater precision. Health information must be normalized and shared between suppliers. HIPAA (Health Insurance Portability and Accountability Act), HITECH (Health Information Technology for Economic and Clinical Health Act of 2009), HL7 (Health Level 7), and other health normalization and monitoring bodies have characterized some standard operational rules [20]. The Authorized Testing and Certifying Body (ATCB) provides outsider opinion on EHR.
2. **Uninterpretable Black Box Model:** Systematic DL opened up new avenues in the area of clinical imaging, prompting new opportunities. DL may solve the unpredictability, which was not possible with conventional AI-based methodologies. One of the greatest barriers is the black box-specific model. The weighted frameworks in neural networks, with incremental depth of layers, makes the model uninterpretable.

4.7 CONCLUSION

DL has undergone significant advances in the medical field, due to the technological revolution that is taking place around the world. First, there was the digitization of medical images and radiological environments. Then, the huge evolution of DL, computer vision techniques, machine learning, and artificial intelligence led to the large-scale development of medical diagnostic systems. More recently, development of DL techniques and computational models has provided great support to the medical decision-making and prognostic prediction processes.

DL algorithms will certainly provide support to reduce the latency of medical examinations, shorten the time required to act in urgent cases, increase diagnostic confidence, and make image analysis and diagnosis more objective and reproducible.

REFERENCES

1. Wedyan, Mohammad, Alessandro Crippa, and Adel Al-Jumaily. "A Novel Virtual Sample Generation Method to Overcome the Small Sample Size Problem in Computer Aided Medical Diagnosing." *Algorithms* 12.8 (2019): 160.
2. Toussie, Danielle, et al. "Clinical and chest radiography features determine patient outcomes in young and middle age adults with COVID-19." Radiology (2020): 201754.
3. Puntmann, Valentina O., et al. "Outcomes of cardiovascular magnetic resonance imaging in patients recently recovered from coronavirus disease 2019 (COVID-19)." *JAMA Cardiology* (2020).
4. Farhadi, Arash, et al. "Ultrasound imaging of gene expression in mammalian cells." *Science* 365.6460 (2019): 1469–1475.
5. Gonzalez-Hernandez, Jose-Luis, et al. "Technology, application and potential of dynamic breast thermography for the detection of breast cancer." *International Journal of Heat and Mass Transfer* 131 (2019): 558–573.
6. Visvikis, Dimitris, et al. "Artificial intelligence, machine (deep) learning and radio (geno) mics: definitions and nuclear medicine imaging applications." *European Journal of Nuclear Medicine and Molecular Imaging* (2019): 1–8.
7. Bashar, Abul. "Survey on evolving deep learning neural network architectures." *Journal of Artificial Intelligence* 1.02 (2019): 73–82.
8. Hudaa, Syihaabul, et al. *Natural Language Processing Utilization in Healthcare.* (2019).
9. Alshehri, Abdulelah S., Rafiqul Gani, and Fengqi You. "Deep learning and knowledge-based methods for computer aided molecular design--toward a unified approach: State-of-the-art and future directions." arXiv preprint arXiv:2005.08968 (2020).
10. Mikołajczyk, Agnieszka, and Michał Grochowski. "Style transfer-based image synthesis as an efficient regularization technique in deep learning." In *2019 24th International Conference on Methods and Models in Automation and Robotics (MMAR).* IEEE (2019).
11. Sahoo, Abhaya Kumar, Chittaranjan Pradhan, and Himansu Das. "Performance evaluation of different machine learning methods and deep-learning based convolutional neural network for health decision making." In Nature Inspired Computing for Data Science. Springer, Cham (2020), pp. 201–212.
12. Nebbia, Giacomo, et al. "Deep learning of sub-regional breast parenchyma in mammograms for localized breast cancer risk prediction." In *Journal of Medical Imaging 2019: Computer-Aided Diagnosis.* Vol. 10950. International Society for Optics and Photonics (2019).
13. Shahnaz, Ayesha, Usman Qamar, and Ayesha Khalid. "Using blockchain for electronic health records." *IEEE Access* 7 (2019): 147782–147795.
14. Shakeel, P. Mohamed, et al. "Neural network based brain tumor detection using wireless infrared imaging sensor." *IEEE Access* 7 (2019): 5577–5588.
15. Tsohou, Aggeliki, et al. "Privacy, security, legal and technology acceptance requirements for a GDPR compliance platform." In *Computers & Security.* Springer, Cham (2019), pp. 204–223.

16. Harrer, Stefan, et al. "Artificial intelligence for clinical trial design." Trends in Pharmacological Sciences (2019).
17. Padala, Suresh, and Veerababu Padala. "Fusion of CT and MRI scanned medical images using image processing." *Journal of Computer Technology & Applications* 3.3 (2019): 17–20.
18. Reynaert, N. "PET and MRI based RT treatment planning: Handling uncertainties." Cancer/Radiothérapie (2019).
19. Stankovic, Zoran, Bradley D. Allen, Julio Garcia Kelly B. Jarvis, Michael Markl. "4D flow imaging with MRI". *Cardiovascular Diagnosis and Therapy* 4.2 (2014): 173–192. doi: 10.3978/j.issn.2223-3652.2014.01.02.
20. Kim, Yejin, et al. "Federated tensor factorization for computational phenotyping." In *Proceedings of the 23rd ACM SIGKDD International Conference on Knowledge Discovery and Data Mining* (2017).

5 Machine Learning Classifiers in Health Care

K. Sambath Kumar and A. Rajendran

CONTENTS

5.1 INTRODUCTION

Machine learning (ML) is widely regarded as one of the most important innovations to have taken place in healthcare systems. ML involves the generation of additional information from clinical data to aid diagnosis, with the use of algorithms and models. Progress by ML is driven by increased access to enormous volumes of clinical data and the ease of calculations. ML centers around calculations based on the machine's past encounters. In basic terms, ML is characterized as the extraction of information from datasets. The objective of ML is to recognize patterns in data and afterward to identify valuable evidence using those examples for which training has been achieved.

The traditional and ML algorithm block diagrams are described in Figures 5.1 and 5.2. ML in healthcare has, of late, been recognized as being truly newsworthy. Google has developed an ML algorithm to detect harmful tumors on mammograms, whereas Stanford University is using an ML algorithm to characterize skin cancer growth, showing that ML provides important tools in the clinical dynamic. ML is different from human interpretation of data on a very basic level, as it has the ability to learn and develop from past experience. Utilizing different programming procedures, ML algorithms can process a great deal of information and extract valuable features and patterns. Along these lines, ML algorithms can simply carry out the tasks they were customized to do or can enhance these interpretive functions by building on the information with which they were supplied. Likewise, there are several types of ML algorithm that are used in certain case studies.

5.1.1 Supervised learning

Most of the valuable ML uses involve supervised learning. In this algorithm, the output (Y) is labeled for all input data (X). It can be expressed as the following:

$$Y = \text{function}\ (X)$$

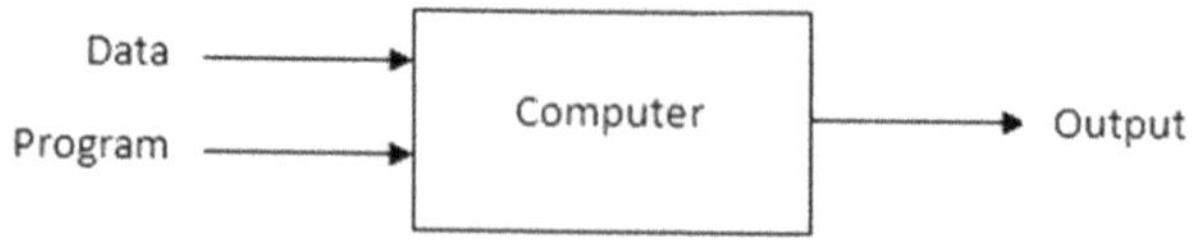

FIGURE 5.1 Traditional programming.

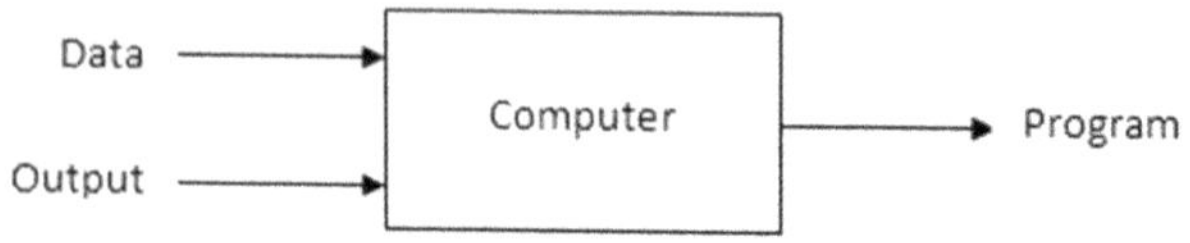

FIGURE 5.2 Machine learning algorithm.

This algorithm generates a trained model from training datasets with the guidance of an instructor. This supervised algorithm continuously predicts the output on the basis of the training input, which is rectified by the instructor. The algorithm is stopped once it accomplishes a reasonable degree of execution. Supervised learning can be sub-categorized into regression and classification.

- **Classification:** The output variable is any one of different mutually exclusive categories, such as "black" or "blue", or "tumor" and "no tumor".
- **Regression:** The relationship is determined between the input data and the output data, the value of which is shown to be real, such as "length" or "price".

Classification problems would include situations where a doctor categorizes patients into healthy or infected, or, in some cases, as having low or high survival rates. The regression problem is also used in the financial market to predict rates of change of stocks.

5.1.2 Unsupervised learning

Unsupervised learning involves having only input data(X) but no labeled outputs. The objective of unsupervised learning is to detect the patterns or features from the given information. Unlike supervised learning, there is no instructor to obtain a specific output for given input data;an algorithm has to find structure or pattern from the given input data. It is again sub-classified into clustering and association rules.

- **Clustering:** To form groupings based on the data, such as to assemble a group based on age.
- **Association:** An association involves finding a relationship or dependency, such as people who purchase X also buy Y, or older people are more at risk from a particular disease.

5.1.3 Semi-supervised learning

In situations where there is a lot of data(X), just a portion of which is labeled(Y) in training, this situation is categorized as semi-supervised learning. This category lies between supervised and unsupervised learning. These unlabeled data are easier to store than labeled data, occupying less memory. Unsupervised learning will achieve similarities with the given input data. Supervised learning is used to find predicted results well. Initially, this algorithm makes a cluster with the available data, after the creation of groupings, with the help of labeled data, remaining unlabeled data to be labeled. More ML problems will be described in this section. Some magnetic resonance imaging (MRI) images are labeled, with the remaining images not being labeled (e.g., normal cells or tumorous).

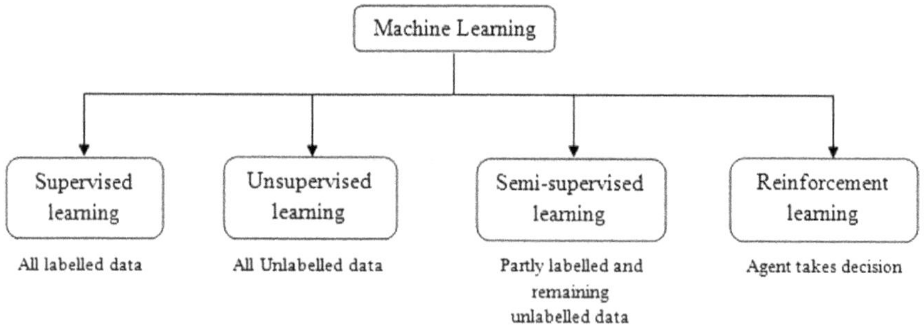

FIGURE 5.3 Types of machine learning.

5.1.4 Reinforcement Learning

Reinforcement learning involves the training of ML models to settle on a sequence of decisions. The specialist figures out how to accomplish an objective in the complex situation in question. In reinforcement learning, an ML faces a game-like situation. The computer uses experimentation to identify an answer for the problem. ML gets either rewards or penalties for the activities it performs. It will likely boost the complete rewards.

In spite of the fact that the designer sets the reward strategy, that is, the guidelines of the game, s/he gives the model no indications or proposals regarding how to play the game. The designer depends upon the model to make sense of how to play out the undertaking to expand the reward, beginning from arbitrary preliminaries and ending with complex methods and abilities superior to those of the humans. By using the intensity of the search and numerous preliminaries, reinforcement learning is presently the best method to exploit the machine's innovation. As opposed to people, ML can develop an understanding from a great many equal ongoing interactions if a reinforcement learning algorithm is run on an adequately ground-breaking computer foundation. The major classifications of ML algorithms [1] are depicted in Figure 5.3.

5.2 DECISION MAKING IN HEALTH CARE

Before ML algorithms can be used in healthcare applications, they need to be 'trained' through data that are created from clinical exercises, e.g., screening, analysis, treatment task, etc., with the goal that they can learn comparable gatherings of subjects, relationships between subject features, and the results of premia. Such clinical information frequently exists, though is not constrained to the type of socioeconomics, clinical notes, electronic data from clinical gadgets, physical assessments, and clinical research center images.

5.2.1 Introduction

Health-related data are monitored by a physician who then takes additional time to examine and interpret the data. When automatic health-related analysis, based on monitoring systems, came into operation, this proved beneficial for society. ML plays a very important role in these healthcare services, which show good results in terms of tumor segmentation, image synthesis, diagnosis, prognosis, etc.

ML has improved data handling in the human services field. It gives a better answer, reduces the cost of healthcare, and provides a good connection between physician and patient. ML algorithms are used in numerous areas of health-related services. Physicians find it easier to recognize problems taking place and will provide medications and, if necessary, will propose treatment planning. These steps are recorded to make follow-ups easier. At present, in healthcare, a huge volume of data has become available. All structured and unstructured data are available in the form of electronic health records (EHR). Structured data is divided into numerous sub-sections for convenience, on the basis of the patient and their health information, such as fever, cold, headache, and so on. Unstructured data are collected in various formats, like reports, images, sounds, video recordings, and prescriptions. These two classes of data (structured and unstructured) improve predictions and increase the effectiveness of treatment results.

5.2.2 Clinical Decision Making

ML tasks have, until recently, been limited to those carried out by humans in tedious, low-level assignments, where pattern recognition or perception are basics. A few models have included data acquisition, normalization, and classification or feature extraction. For more significant levels, including interpretation, for example, patient's status interpretation and help with decision making, ML takes into account the combination of decision making [2] and handling of heterogeneous information in the dynamic procedure, yet these are still formal and need considerable approval. Figure 5.4, which outlines the way toward settling the clinical decision-making process, demonstrates how ML could contribute. Figure 5.4 also outlines that the risk to a patient from mistakes increases incrementally with each progression. Diverse ML approaches can complement the work of clinicians in various assignments, helping in particular in terms of pattern recognitions where complex information is considered, yet they should be interpreted and directed by clinicians [3].

Data acquisition is the principal phase of the dynamic flowchart. It is critical as the extracted information will depend on the inherent quality and measure of the accessible information. In a perfect world, this information should highlight the condition under investigation, particularly the presence of pertinent anomalies. To arrive at a good understanding of most ailments, a lot of information is required; the more information, the better.

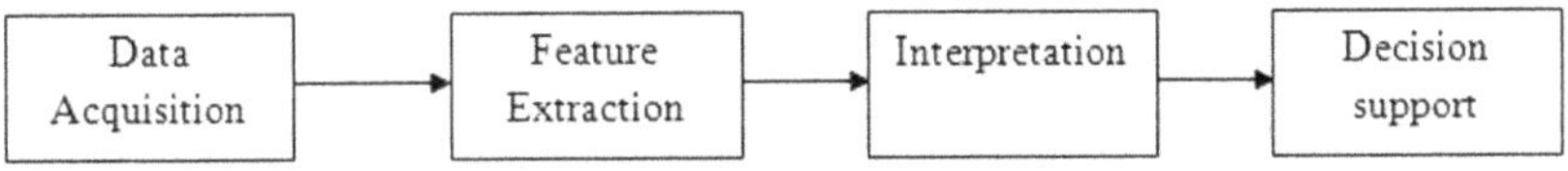

FIGURE 5.4 Steps in the clinical decision-making process.

Once the data are acquired and curated (i.e., standardized and organized), the next stage toward improved clinical decisionmaking is to extract features to obtain a better understanding. ML techniques enable the extraction of relevant features from data such as signals, laboratory results, or images, among others. In imaging, for example, these features can be geometrical, functional, or based on local voxel intensity or texture, and relate to an anatomical region (organ or body substructure), the detection of which is automatically achieved by segmentation, currently the most common subject of ML papers in medical imaging like Wolterink et al. and Dalca et al. employed a convolutional neural network (CNN) to segment tumor in brain MRI and the coronary arteries centerline extraction in cardiac computed tomography (CT) angiography [4, 5].

When the clinical information on a patient is obtained and relevant features are extracted, the next stage in the decision-making procedure consists of deciphering the patient's status by correlation with others from an unhealthy or control population. Data normalization is required for inter-individual comparison. At the point where complex information is included, for example brain or heart images, the customary way to deal with normalization is to construct a factual map book, a reference model that characterizes the variability related to the training population. To construct an atlas, the training data must be changed into a typical spatial structure, which can be accomplished by getting number of features. In this sense, registration takes into consideration the combined and effective examination of population information, a vital advance for interpretation of patient status toward diagnosis.

Registration is another field where ML has been shown to be useful. Two methodologies are commonly used: 1) utilizing ML structures to obtain a similarity measure in order to drive an iterative improvement procedure from which to decide the ideal change boundaries; or 2) utilizing regression networks to legitimately predict the change boundaries. Unsupervised learning is used for 3D mind registration. Dalca et al. [6] built up a learning system for clinical image registration that is intended to accelerate registration time while keeping the precision from being affected by deformation models.

In light of the interpretation of the status of the patient, clinicians needs to make a decision, which comprises of: 1) watching the patient and delaying until an occasion triggers the requirement to make a decision; 2) gathering more information to improve the chances of taking an appropriate decision; or 3) playing out a mediation, trailed by checking to assess the result of the treatment. AI techniques can assist the clinician with deciding which pathway to follow, in a way that is also financially sound. Making a decision is maybe the most problematic phase of the dynamic pipeline, as endorsing a particular treatment infers realizing the risk related with every patient.

5.3 MACHINE LEARNING IN HEALTH CARE

Nowadays, ML provides numerous applications in the healthcare domain, which includes patient's data collection, early diagnosis, patient monitoring, and treatment planning [7].

5.3.1 Introduction

The healthcare management system, similar to the arrangement of public health or medical care, is basically a cross-section of data handling undertakings. Strategy creators adjust the healthcare system elements of organization and administration, financial and resource management to accomplish healthcare system outputs and system objectives.

The arrangement of medical services includes two tasks: firstly, screening and conclusion, which is the grouping of cases dependent on the history, assessment, and investigations, and secondly, therapy and monitoring, which includes the planning, implementation and observing of a multi-step cycle to convey a future result. The basic type of these cycles over the areas of healthcare system management and the arrangement of care includes data generation, hypothesis testing, and action. AI can possibly improve data generation and hypothesis testing inside a healthcare management system by revealing patterns in the information, and, subsequently, has the potential for considerable beneficial effects on both the individual patient and at the system level.

ML develops existing measurable procedures, using strategies that are not based on earlier presumptions about the delivery of the information, and can discover patterns in the information that can therefore be utilized to achieve data generation and allow hypothesis testing. In this manner, while ML models are harder to interpret, they can consolidate a lot more factors and are generalizable over a more extensive cluster of information types, creating results under difficult circumstances. These kinds of organizations and in most cases are dependent on information from a single source, with greater potential for reproducibility and generalizability [8]. However, the rapid progress of ML takes place inside both the medical services and all data handling undertakings in the public arena.

The developing techniques like ML [9] provide better information to physicians about patient health. ML reduces clinical costs, like routine laboratory tests, and the time to diagnosis. Generally, in healthcare, the huge dataset is trained for a particular disease, after which new patient data are tested, based on that trained model (if its training accuracy is accepted by the physician). Then, based on that testing report, the treatment plan is taken care of by the physician. ML is capable of giving better prediction results than human physicians from a large dataset, and converts information with medical insight knowledge to guide the physician to develop a treatment plan. These arrangements provide good output results, in less time and at less expense.

5.3.2 Opportunities for ML in Health Care

Medical care specialist organizations generate a lot of heterogeneous data, making it difficult for "conventional techniques" to analyze and measure. ML techniques help to successfully break down this information into noteworthy pieces of knowledge. Furthermore, there are heterogeneous information sources that can increase medical services information, such as genomics, clinical information, information from online media, and so on. The four significant applications of medical services that can profit from ML procedures are prognosis, diagnosis, treatment, and clinical work process.

Prognosis is the strategy for anticipating the normal improvement of an infection in clinical practice. It also incorporates identification of symptoms associated with a particular infection, determining whether ailments will turn out to be worse, improve, or stay stable over the long run, and achieving recognizable proof of potential related medical issues, difficulties, capacity to perform routine exercises, and the probability of endurance. As in the clinical setting, multi-modular information is gathered from patients, which can be utilized by ML models, with the results then being passed on to the physician to adopt necessary treatment planning, assess the complexity of the affected organ area, and identify the medical care given to the patient. For example, ML models have been largely created for the identification and grouping of various types of cancers, e.g., brain tumors.

In any case, the potential applications of ML for infection prognosis, i.e., expectation of illness side effects, risks, survivability, and recurrence, have been misused under ongoing translational examination approaches that plan to achieve customized medication. Nevertheless, the field of customized medication is beginning, requiring broad improvement in associated fields like bioinformatics, solid approval methodologies, and obviously powerful uses of ML to accomplish this huge and translational effect.

In medical image analysis, ML procedures are utilized to achieve efficient and useful extraction of data from clinical images from methods which use distinctive imaging modalities, for example, X-ray, angiography, MRI, fluoroscopy, CT, nuclear medicine, mammography, and so on. These modalities give significant amounts of useful data about various body organs and will play a critical role in the detection and determination of abnormalities. The critical motivation behind clinical image analysis is to help physicians and radiologists achieve effective analysis and anticipation of the conditions. The important tasks in clinical image investigation involve detection, clustering, classification, segmentation, reconstruction, and image registration. Additionally, completely autonomous frameworks for smart clinical image diagnosis are important for cutting-edge medical service systems.

Medical images are generally employed in routine clinical practice and the investigation and interpretation of these images are performed by doctors and radiologists. To reveal the discoveries with respect to the images under consideration, textual radiology reports are composed about each body organ that is inspected in the direct examination. Composing such report is exceptionally testing under certain situations, e.g., when undertaken by less experienced radiologists and medical care specialist organizations in provincial regions, where the nature of medical services available is inadequate. On the other hand, for experienced radiologists and pathologists, the generation of high-quality reports can be repetitive and tedious, a problem which can be exacerbated by the large number of patients being seen every day.

The above examples are ML-based healthcare applications developed by companies to benefit patients and healthcare providers. Figure 5.5 describes the healthcare

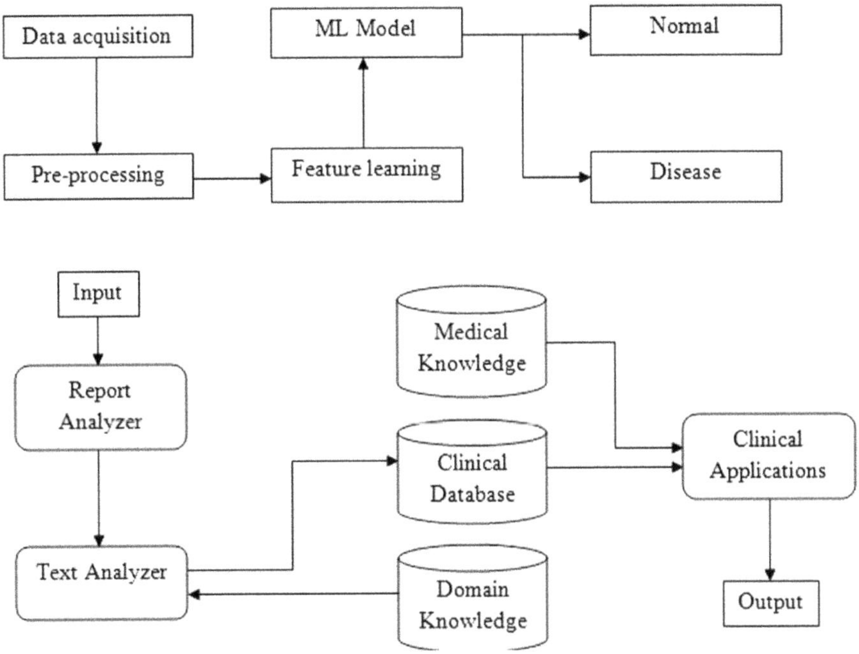

FIGURE 5.5 Healthcare applications developed by different companies.

opportunities realized by some companies. A multi-tasking ML-based system is also proposed for automatic labeling and presentation of medical images.

5.4 DATA CLASSIFICATION TECHNIQUES IN HEALTH CARE

Healthcare is a serious industry where patients and medical organizations are dependent upon critical administrative oversight. The complex needs of these associations must be delivered in as effective and accurate a way as possible under the circumstances, while avoiding potential risk, in order to guarantee appropriate degrees of data security and privacy.

Given the data involved, and the extremely intimate relationship of such data to their patients, those in the healthcare industry have a clear understanding that the data they depend on is central to patient care – any loss of accuracy can have enormous negative effects. Whereas they are not regarded as an element of 'data governance', healthcare service professionals have an appreciation for the significance of the data and the requirement for its protection and accessibility, as they utilize the data that ultimately drives the decisions that they make throughout the day. With obligations toward consideration of past patients, there are security concerns. Like all data-driven associations, the pivotal need to create and actualize data governance capacities has never been more prominent, as the data that drive our understanding are additionally the source of legal and business risk.

5.4.1 Support Vector Machine

The support vector machine (SVM) is the preferred data mining method when the test dataset is small and includes many features. SVM is important for classification problems. The SVM algorithm prepares the hyperplane for the given training dataset to make one of the two classifications. This method categorizes the given training data accurately with the help of the hyperplane, which is used to separate two different data clusters efficiently.

In practice, however, it can be unrealistic to clearly isolate the given dataset since a portion of the information focus in the two classes may fall into a disease-defined situation that is difficult to isolate directly. To overcome this problem, higher measurement data are prepared, and, at the end, two different classes can be distinguished by the higher measurement. Also, as referenced above, most training datasets used in an assortment of areas are not directly isolated and it is difficult to conclude the effective ideal hyperplane to make the two class vectors. A few arrangements called part works have been proposed and reviewed to overcome these non-linearity problems for SVM.

Very-large-scale datasets, missing or flawed features, and skewed transport of classes are basic difficulties in design associated with numerous issues of healthcare services. The effective expansion of a ground-breaking ML procedure, support vector machines, to the versatile staggered system of a cost-effective learning SVM enabled management of imbalanced classification problems. The staggered system considerably shortens the computational time necessary without losing the nature of the classifiers for extremely large datasets. Son et al. [10] indicated that the multi-level weighted support vector machine (MLWSVM) produces more considerable outcomes than do the multi-level support vector machine (MLSVM) and the standard SVM methods as a rule. This work can be extended to handle other classification problems with enormous volumes of unbalanced data (consolidated from various sources) with missing features in healthcare and building applications.

The drawn-out human services role of this kind of work comprises of two sections: to show general points of interest for applying a particular AI strategy to deal with healthcare data, and to contend with the employment of a multi-scale portrayal of complex clinical information, coordinated from different sources and containing uncommon events. In spite of the fact that the outcomes introduced here are not perfect (perhaps because of the multifaceted nature of the features), they are adequate to suggest the value of this strategy for future use in predictive analysis in healthcare.

5.4.2 Logistic Regression

Logistic regression is a method from statistics, which is used for binary classification problems. Sometimes it can be difficult to interpret [11]. The health of a kidney transplant patient was monitored with a logistic regression model [12]. In this model, three different groups of people were selected. Out of these, 70 healthy people and 136 kidney transplant patients were taken to represent the training dataset, whereas the validation dataset involved110 kidney transplant patients, with the 36-Item Short Form Survey (SF-36) score, an oft-used, well-researched, self-reported measure of health, being used to measure the healthy group. The derived model was developed by the training dataset and this model was tested with the validation dataset. From this model, metric factors like sensitivity, specificity, and accuracy were calculated [12].

5.4.3 Artificial Neural Network

An artificial neural network (ANN) is a supervised algorithm which is consists of more number of neurons. For the most part, an ANN includes countless processors working equally and organized in levels. At the beginning level of this algorithm, the raw data are given, which is the same as a human visual system. The next level input is taken from the previous level, which will be continued to the last level of this algorithm. This model output is taken from the previous level. This improved technology is utilized by a greater number of sectors, such as the financial sector, the healthcare domain, remote sensing, and so on. This algorithm has several applications to the financial sectors, like recognition of handwritten documents, speech-to-text conversion, facial recognition, and biometric recognition.

With the digitization of healthcare services, clinics are becoming ready to gather a lot of data across huge data frameworks. With its capacity to process huge datasets, ML innovation is appropriate for dissecting clinical information and developing successful algorithms. Considering the common use of clinical data frameworks and clinical databases, ANN has a greater number of healthcare applications.

In ANN, the back-propagation algorithm plays an important role in classifying outputs accurately, by modifying weights with help of a feedback path. The absence of straightforwardness or interpretability of neural networks keeps on being a significant problem since human services suppliers are regularly reluctant to accept machine suggestions without clarity with regard to the basic method of reasoning. Handling healthcare applications, such as biomedical signal processing, speech processing, health report generation, and follow-up data generation was complex before 2000 but, after the introduction of deep learning (DL), it became easier with a very difficult structure, like the ImageNet Large-Scale Visual Recognition Challenge, which classifies more than one billion images in record-breaking time.

ANN is used in more medical applications, such as complex brain structure segmentation. ANN is implemented for diagnosis, classification, prognosis, segmentation, and prediction of the growth of the affected organ. An ANN algorithm implemented for healthcare decision making at all levels by Shahid et al [13]. It has four zones, namely the detection and treatment plan, electrocardiograph (ECG) signal, and detection of the cause of the problem from heart images and medical care. Teleservices offer remote services, which tend to monitor, indicate, report generation, take follow-ups, analyze, plan treatments and care after treatment. The following examples use ML techniques, especially ANN, to handle frequent blood pressure (BP) measures, grouped into high-risk and low-risk patients, and with medical report generation from ECG signals.

5.4.4 Random Forest

The Random Forest(RF) technique involves large numbers of tree and belongs to the category of supervised learning algorithms. It deals with both classification and regression type problems, although it is primarily usedfor classification problems. Additionally, RF predicts the best-choice tree among a number of options. It is grouping method, compared with the Decision Tree (DT) method, because it reduces overfitting problems by taking the average output.

As enormous human services datasets increase, irregular tree models will almost certainly be progressively more commonly used to address inquiries of clinical interest. Irregular RF models have been employed, for instance, in the medication improvement process, in genomic information analysis, and using imaging information to predict results in the case of Alzheimer's disease. Arbitrary RF and other ML approaches will be largely adjusted to concentrate on enormous number of input factors using huge datasets, and to increase accessibility with respect to medication in the future.

Data: MRI images were taken of 33 patients, with all patients' images from each patient taken of the organs, namely head, heart, liver, neck, and lungs [14]. Detection of organs uses the water and fat channels inside the human body. For all organs, 3D images were taken, with pixel dimensions of 192, 144, and 443 for height, width, and depth, respectively.

Regression approach: In the training phase, 20 patient MRI images were taken, and for the testing phase, MRI images were taken from the remaining 13 patients[14]. For the Random Forests and Random Ferns approaches, all parameters, such as the number of trees have been tuned and recorded. For faster training and evaluation, Random Ferns achieved very good localization accuracy. To conclude, detection and localization of organs provided by this proposed Random forest and Random fern approaches was better than with the existing approaches.

5.4.5 Decision Tree

Decision Tree (DT) is the very powerful and well-known tool for classification and prediction. It is a flowchart-like tree pattern, with a test on an attribute denoted as an internal node, an outcome of the test described by each branch, and with each terminal node given a class label. The source dataset can be partitioned into subsets by training. In a recursive manner, the above process is repeated to derive a subset. This process is stopped when the predicted value is optimized or when all main nodes provide the same prediction value. The development of the DT classifier does not require any domain knowledge or parameter setting. Datasets with more features can also be handled by DT.

In this study by Chern et al.[15], to classify the possibility of achieving Telehealth administrations through health insurance reimbursements, a heuristic DTTelehealth classification approach (HDTTCA), can be used, which comprises of three significant advances, namely (1) analysis from the dataset, (2) the DT model, and (3) the predicted output, as shown in Figure 5.6 [15]. This HDTTCA is used for Telehealth

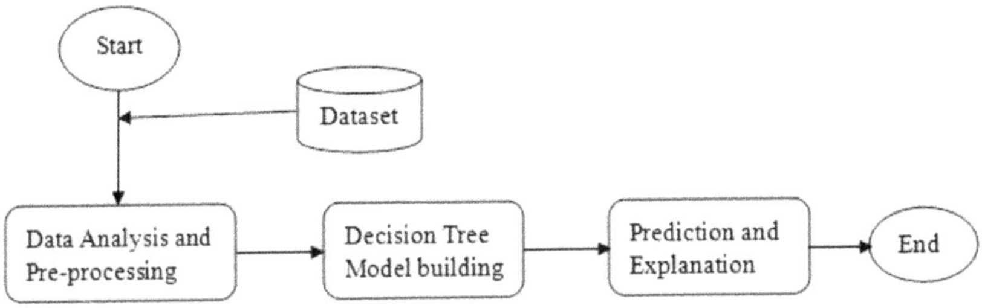

FIGURE 5.6 Heuristic Decision Tree Telehealth classification Approach (HDTTCA) flow chart.

services. In the wake of finding the attainable combination of variables, models, and related boundaries in the Method [15], testing the results with the logical regression to make this model even better, the Telehealth classification result is better for the patients.

5.4.6 *K*-Nearest Neighbor

K-Nearest Neighbor (KNN) is a straightforward algorithm that has different numbers of cases and groups, with the new data being case dependent on a similarity measure. It is generally used to classify data, listing a data point relative to how its neighbors are ordered. In the study by Sun et al.[16], a chronic kidney dataset from the UCI Machine Learning Repository was used, this dataset containing information related to kidney infection on 400 patients, measuring25 different parameters over a2-month time period. Every one of the 25 parameters contained in the dataset is used in this KNN algorithm. The study also classifies heart disease from a heart disease dataset collected from the UCI Machine Learning Repository, with this information being gathered from the Cleveland, Hungazerland and the VA Long Beach datasets.

The classification result for the two above-mentioned datasets was analyzed as described by Sun et al. [16] and its accuracy was estimated to be 90% approximately[16]. *K* represents the number of neighbors and it is always an odd number. For each *K* value, this model is evaluated 100 times (epochs). When the *K* value is less than 5, then the accuracy varied, reducing the accuracy of the derived model. When the *K* value is more than 5 but less than 13, then accuracy can be achieved consistently from this derived model.

5.4.7 Naïve Bayes

The Naïve Bayes method is a classification technique which follows the Bayes' theorem. In straightforward terms, a Naïve Bayes classifier collect features from particular groups, which will vary from other features. Likewise, independent features are gathered by using this classifier. To detect an apple, the following features are considered: 3 inches in diameter, round in shape, and red in color. Some other features are also available, which are also included to detect an apple fruit. This is the specialty of the Naïve Bayes algorithm. This method is suitable for huge dataset collections which perform well in highly sophisticated classification problems.

Of the articles in this examination, 21.7% included brain diseases, such as brain tumor, metastasis, trauma, and Alzheimer's disease. CT is broadly acknowledged to be a successful method to analyze and recognize uncommon yet clinically important features in patients suffering a minor head injury or a brain infection [17]. In that capacity, CT has been increasingly used as a straightforward and accessible test for these patients. In any case, a fundamental report directed by Brenner and Hall [18] cautions against its negative impacts (especially for children) because of the introduction of radiation with CT. Independent CT imaging advocates the appropriation of a far-reaching approach that objectivizes doctors' instructions to reduce the overdependence on CT imaging for those experiencing brain disease. Accordingly, to reduce the destructiveness of CT, it is considered to be useful to employ data-mining methods, with the neuron-by-neuron (NBN) algorithm as a predictor.

5.5 CASE STUDIES

ML quickens the pace of logical disclosure across fields, of which the medical field is no difference. From language handling software, that quickens exploration, to predictive algorithms that alert clinical staff of an impending heart attack episode, ML supplements human understanding and practice across medical disciplines [19].

In any case, with all the "solutionism" around ML innovations, human services suppliers are justifiably concerned about how it will truly support patients. Numerous ML classifiers available for human services objects are customized to take care of specific problems, such as recognizing the risk of developing sepsis or diagnosing breast cancer. These state-of-the-art ML classifiers make it difficult or even impossible for organizations to re-cast their models and to capitalize on their investment.

Open-source data science permits healthcare firms to adjust models to address a variety of problems, utilizing the most recent ML advances, such as sound and vision data processing. Employing open-source tools, data researchers can customize applications such that they meet human services and improve applications in a variety of settings, at last differentiating an organization from its competitors.

5.5.1 Brain Tumor Classification

Brain cancer is the main cause of death globally, as indicated by the World Health Organization (WHO). Early detection of cancer can delay death, although early detection is generally not possible. A tumor can be benign, pre-cancerous, or malignant. Benign tumors differ from malignant tumors, as they do not spread to different organs and tissues and can be carefully removed. Examples of brain tumors are gliomas, meningiomas, and pituitary tumors. Gliomas are tumors that develop from the brain cells, unlike meningiomas. Meningiomas emerge from the layers that cover the brain and encompass the central nervous system, whereas pituitary tumors are anomalies that sit inside the skull. The most significant difference between these three kinds of tumors is that meningiomas are ordinarily benign, whereas gliomas

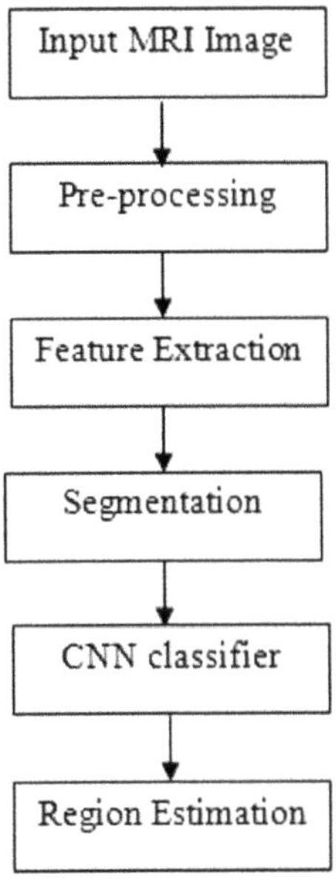

FIGURE 5.7 Flowchart of architecture.

are usually malignant. Pituitary tumors, regardless of whether they are benign or not, can cause other clinical damage, in contrast to meningiomas, which are slow-developing tumors. In view of the data referenced here, the exact separation between these three kinds of tumors requires significant advances in the clinical analytic procedure and subsequent evaluation of the patient. The flow of the brain tumor segmentation process [20] is explained with help of Figure 5.7.

5.5.1.1 MRI Brain Image Acquisitions

In a survey of brain tumor segmentation techniques, Kumar and Rajendran [21] developed a ML model for MRI images from 220 HGG (high-grade glioma) and 54 LGG (low-grade glioma) patients, taken from the BRATS 2015 training dataset. Each patient was imaged by multimodal MRI, containing four sequences, namely T1, T1C, T2, and FLAIR [21].

5.5.1.2 Preprocessing

Before inputting the raw pixels to the trained model, any artifacts will be removed. This task is carried out by preprocessing techniques, like normalization, histogram equalization, and so on. MRI sometimes has a bias which affects segmentation results, especially in an automatic segmentation model. The Chmelik et al. study used the n4T1K preprocessing technique to remove bias from MRI images [22]. Furthermore, noise level is reduced by using a median filter, which normalizes pixels in an image. The above two preprocessing techniques subsequently reduced noise level and removed artifacts from an image, which improving segmentation results.

5.5.1.3 Convolutional Neural Network (CNN) Algorithm

The CNN algorithm plays out the voxel-wise classification problem. The tumor or affected area is isolated by voxel-wise comparison of images. The CNN algorithm has four section input and convolution areas. In an image convolution patch added after convolution, the resultant image is reduced. Depending on the number of layers and the padding type, input image size is further reduced. Finally, from the last output layer, all features are combined, and a feature map is generated. The prediction score is generated from the classification section.

5.5.1.4 Training of the Network

The training model is implemented with Keras, an open-source software library that provides a Python interface for ANNs. The sequential model consists of four layers, trained on MRI images of 33× 33 patches to achieve classification of the center pixel. Four channels are available in each patient's MRI. From these four channels, an imaging sequence is identified by one channel and the net gain in proficiency is reflected by the relative pixel intensity of each given class.

5.5.1.5 Validation of Data Set

Validation results are measured with help of the BRATS 2015 validation dataset. The validation dataset also contains four different modalities, namely T1, T1C, T2, and FLAIR.

5.5.1.6 Results

The segmentation result had four different labels: edema, advanced tumor, non-advanced tumor, and necrotic core, which are assigned label 1, label 2, label 3, and label 4, respectively. The model has a height of 220 pixels and a width of 220 pixels, which is suited for color image or gray images [22]. Automatic brain tumor segmentation consists of two main tasks, segmentation and edge detection.

The segmentation task compares the pixels of the normal brain image and the abnormal brain image. From this comparison, abnormal tissues are localized by this specific task. The next task of edge detection is performed with leading edge detection by a Canny edge detector, which isolates brain tissues from segmented images. This output image shows the tumor colored white. In Figure 5.8, the red color represents the core tumor, blue color is the advanced tumor, and green color is edema.

5.5.2 Breast Cancer Classification

Breast cancer is a very serious disease which occurs largely in women, where breast cells develop in an uncontrollable manner. Patients need to know about side effects that include another irregularity, growths in the surrounding breast cells, irritation, and discharges other than milk. When a doctor finds abnormalities in the breast, a biopsy is generally recommended. By this methodology, breast tissue is collected and analyzed under a microscope. After confirming breast cancer, the clinical group will classify a treatment plan. To achieve selection of the correct

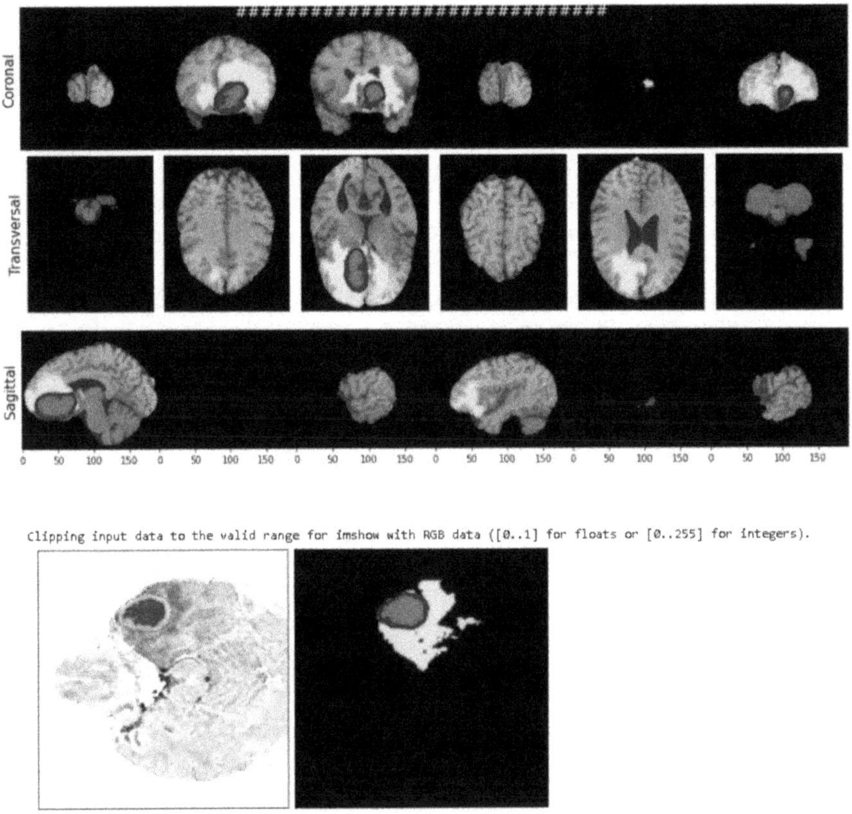

FIGURE 5.8 Brain tumor segmentation results.

treatment, doctors characterize the individual breast cancer, as indicated by standard boundaries.

Dabeer et al.[23] presented and surveyed an ML algorithm for automatic breast cancer recognition that combines elements of ML detection and image classification. The input image from the raw pixels highlighted the visual examples, and afterward uses those examples to distinguish non-cancerous from cancerous tissue, which highlights image segments critical for making decisions, with the assistance of a classifier. The CNN was trained using 2480 benign and 5429 malignant examples with the red, green and blue (RGB) shading model. In this way, the proposed framework portrayed in Figure 5.9 provides an effective classification model by which to classify breast tumor tissue as being either benign or malignant. The process of classification is done by taking the flattened weighted feature map obtained from the final pooling layer and using it as input to the fully connected layer, which calculates the loss and modifies the weights of the internal hidden nodes accordingly. These layers are stacked after preprocessing is completed. The output of the last layer is taken as the final output, as usual. A training accuracy of 93.45% was obtained with a test train split of 0.2. Of the evaluation metrics, precision is a probabilistic measure to determine whether a positive case, defined by the authors, actually belongs to the positive class. A recall is a probabilistic measure to determine whether an actual positive case is correctly classified with the positive class. The F1 score is calculated as the geometric mean between precision and recall.

The precision outcomes for the benign and malignant classes are shown as 90.55% and 94.66%, respectively, with the recall values for the two classes being 87.66% and 95.98%, respectively [23].

From the aforementioned classification problems, the degree of accuracy was improved relative to the state-of-the-art experimental setups. Alongside this improvement in accuracy, there was a noteworthy improvement in precision and recall. This methodology is helpful as this model is completely automated and any user can test another image just by selecting it. Indeed, even at the planning stage, there is no requirement for domain knowledge, and it gives high predictions of accuracy. Malathi and Sinthia [18] tested their proposed model with different goals of histopathology images and the outcomes were generally insensitive toward the goal. By employing this computerized algorithm, there is a chance of developing an economical system for recognition of cancer in the early phases, which can eventually increase survival rates among breast cancer patients.

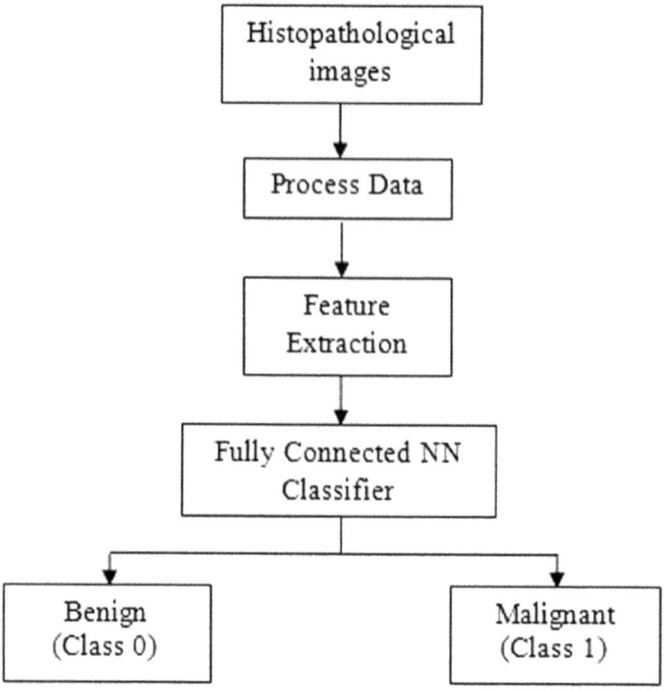

FIGURE 5.9 Flowchart of the breast cancer classification model.

5.5.3 Classification of Chronic Kidney Disease

In recent years, kidney disease has become common in adults, affecting 13% of adults around the world. Chronic kidney disease (CKD) can be categorized into five stages, based on the structure and function of the kidneys. Blood test and urine test samples are taken to evaluate CKD [24]. Both blood and urine tests are performed to get an accurate result with respect to CKD and to project a treatment plan. The high risk of CKD affects heart-related disease, so the high-risk CKD patient undergoes dialysis treatment.

Thomas, [24] discussed self-management, the main issue of which is whether self-management is helpful. One recent survey focused on the basic examination of CKD in stage 1–4 adults, and evaluated whether these mediations improved health proficiency, self-viability, and health-related personal satisfaction, as well as hospitalizations. In spite of the fact that the effectiveness of self-management programs in CKD was not demonstrated, it is attractive that CKD adults are able to self-manage every part of their health.

5.5.4 Classification of COVID-19

Late in 2019, the COVID-19 disease arose in the Wuhan area of China. Early-stage symptoms of this disease are a dry cough, fever, and tiredness. Diagnosis is carried out with the help of a CT scan image of the chest. Serious symptoms of this disease are breathing difficulty, chest pain, and an inability to speak or move. This disease spreads from person to person by droplets from the infected person, by way of coughing or sneezing. Emergency care needs to be given to affected people. At the early stage of infection, the physician takes a CT scan of the lungs for diagnostic purposes [25]. With the advances in computer vision technology, the acquired image is pre- processed before being inputted into the ML model constructed. These pre-processing steps consume time to classify images accurately and to predict the new image. Ozsahin et al. [26] took stomach CT scan images of 150 patients and investigated them at different patch levels of four different subsets. From these outputs, positive and negative cases could be distinguished.

This study carried out a coronavirus grouping in two phases [26]. In the initial stage, the classification procedure has four unique subsets. These subsets are converted into vectors to allow inputting to a SVM. For this model, five feature extraction methods were developed, namely Gray Level Co-occurrence Matrix (GLCM), Local Directional Patterns (LDP), Gray Level Run Length Matrix (GLRLM), Gray Level Size Zone Matrix (GLSZM), and Discrete Wavelet Transform (DWT). Cross-validation methods at the 2-fold, 5-fold, and 10-fold levels were employed by this method. The mean values for classification results were obtained. Figure 5.10 shows the two phases of the classification process. In this study, acquired images were taken from different CT scan tools. Five distinct feature extraction methods isolated contaminated patches from the normal patch, providing accuracy values of 99.68% when classifying the abnormal image.

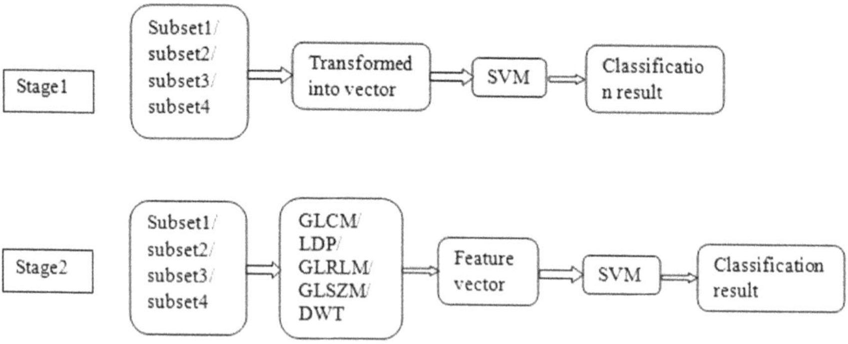

FIGURE 5.10 The stage 1 and stage 2 COVID-19 classification process.

5.6 CONCLUSION

In this chapter, various data classifiers in ML are discussed for the healthcare domain. Therefore, many opportunities are available for ML researchers to collaborate with physicians and medical diagnostic centres, which will encourage machine learning researchers to identify and tackle problems early on with clinical experts. These efforts will develop a trained model, that is useful and feasible for healthcare personnel.

REFERENCES

1. Alić, B., Gurbeta, L., Badnjević, A. Machine learning techniques for classification of diabetes and cardiovascular diseases. In *2017 6th Mediterranean Conference on Embedded Computing (MECO), Bar* (2017), pp. 1–4. doi: 10.1109/MECO.2017.7977152.
2. Wang, Y., Zhang, X., Wang, T. et al. A machine learning framework for accurately recognizing circular RNAs for clinical decision-supporting. *BMC Medical Informatics and Decision Making* vol. 20 (2020): 137. doi: 10.1186/s12911-020-1117-0.
3. Ghassemi, M., Naumann, T., Schulam, P., Beam, A.L., Chen, I.Y., Ranganath, R. A review of challenges and opportunities in machine learning for health. amia joint summits on translational science proceedings. *AMIA Joint Summits on Translational Science* vol. 2020 (2020): 191–200.
4. Havaei, M. et al. "Brain tumor segmentation with deep neural networks." *Medical Image Analysis*, vol. 35 (2017): 18–31. doi: 10.1016/j.media.2016.05.004.
5. Wolterink, J.M., van Hamersvelt, R.W., Viergever, M.A., Leiner, T., Išgum, I. Coronary artery centerline extraction in cardiac CT angiography using a CNN-based orientation classifier. *Medical Image Analysis* vol.51 (2019): 46–60. doi: 10.1016/j.media.2018.10.005.
6. Dalca, A.V. et al. "Unsupervised Learning of Probabilistic Diffeomorphic Registration for Images and Surfaces." *Medical Image Analysis*57 (2019): 226–236. doi: 10.1016/j.media.2019.07.006.
7. Bica, I., Alaa, A.M., Lambert, C., van der Schaar, M. From real-world patient data to individualized treatment effects using machine learning: Current and future methods to address underlying challenges. *Clinical Pharmacology & Therapeutics* vol. 109 (2020): 87–100. doi: 10.1002/cpt.1907.

8. Panch, T., Szolovits, P., Atun, R. Artificial intelligence, machine learning and health systems. *Journal of Global Health* vol. 8, no. 2 (2018): 020303. doi: 10.7189/jogh.08.020303.
9. Sidey-Gibbons, J., Sidey-Gibbons, C. Machine learning in medicine: a practical introduction. *BMC Medical Research Methodology* vol. 19 (2019): 64. doi: 10.1186/s12874-019-0681-4.
10. Son, Y.-J. et al. Application of support vector machine for prediction of medication adherence in heart failure patients. *Healthcare Informatics Research* vol. 16, no. 4 (2010): 253–259. doi: 10.4258/hir.2010.16.4.253.
11. Zekic-Susac, M., Sarlija, N., Has, A., Bilandzic, A. Predicting company growth using logistic regression and neural networks. *Croatian Operational Research Review* vol. 7 (2016): 229–248. doi: 10.17535/crorr.2016.0016.
12. Lavie, P., Herer, P., Hoffstein, V. Obstructive sleep apnoea syndrome as a risk factor for hypertension: population study. *BMJ* vol. 320, no. 7233 (2000): 479–482. doi:10.1136/bmj.320.7233.479.
13. Shahid, N., Rappon, T., Berta, W. Applications of artificial neural networks in health care organizational decision-making: A scoping review. *PLOS ONE*vol. 14, no. 2 (2019): e0212356. doi: 10.1371/journal.pone.0212356.
14. Pauly, O. BOOK: Random Forests for Medical Applications, https://books.google.co.in/books?id=66KgngEACAAJ. Universitätsbibliothek der TU München, 2012.
15. Chern, C.C., Chen, Y.J., Hsiao, B. Decision tree–based classifier in providing telehealth service. *BMC Medical Informatics and Decision Making* vol. 19 (2019): 104. doi: 10.1186/s12911-019-0825-9.
16. Sun, J., Hall, K., Chang, A., Li, J., Song, C., Chauhan, A., Ferra, M., Sager, T., Tayeb, S. Predicting medical conditions using k-nearest neighbors. *Year* vol. 1 (2017): 3897–3903, doi: 10.1109/BigData.2017.8258395.
17. Langarizadeh, M., Moghbeli, F. "Applying Naive bayesian networks to disease prediction: A systematic review." Acta Informatica Medica: AIM: Journal of the Society for Medical Informatics of Bosnia & Herzegovina: casopis Drustva za medicinsku informatiku BiH vol. 24, no. 5 (2016): 364–369. doi: 10.5455/aim.2016.24.364-369.
18. Brenner, D.J., Hall, E.J. Computed tomography--an increasing source of radiation exposure. *New England Journal of Medicine* vol. 357, no. 22 (2007 Nov 29): 2277–2284. doi: 10.1056/NEJMra072149. PMID: 18046031.
19. MathurP. Case studies in healthcare AI. In *Machine Learning Applications Using Python*. Apress, Berkeley, CA (2019), vol. XVIII, p. 379. doi: 10.1007/978-1-4842-3787-8.
20. Malathi, M, Sinthia, P. "Brain tumour segmentation using convolutional neural network with tensor flow." *Asian Pacific Journal of Cancer Prevention: APJCP* vol. 20, no. 7 (2019): 2095–2101. doi: 10.31557/APJCP.2019.20.7.2095.
21. Kumar, K., Rajendran, A. A survey on brain tumor segmentation techniques. *Indian Journal of Public Health Research & Development*vol. 10 (2019): 1070. 10.5958/0976-5506.2019.00439.X.
22. Chmelik, J., Jakubicek, R., Walek, P., Jan, J., Ourednicek, P., Lambert, L., Amadori, E., Gavelli, G. Deep convolutional neural network-based segmentation and classification of difficult to define metastatic spinal lesions in 3D CT data. *Medical Image Analysis* vol. 49 (2018): 76–88. doi: 10.1016/j.media.2018.07.008.
23. Dabeer, S., Khan, M.M., Islam, S. Cancer diagnosis in histopathological image: CNN based approach. *Informatics in Medicine Unlocked* vol. 16 (2019): 100231. doi: 10.1016/j.imu.2019.100231.
24. Thomas, N. Classification of chronic kidney disease 10 years on: what have we learnt and what do we need to do now?*Family Practice* vol. 35, no. 4 (2018): 349–351. doi: 10.1093/fampra/cmy015.

25. Panwar, H., Gupta, P.K., Siddiqui, M.K., Morales-Menendez, R., Singh, V. Application of deep learning for fast detection of COVID-19 in X-Rays using nCOVnet. *Chaos, Solitons, and Fractals* vol. 138 (2020): 109944. doi: 10.1016/j.chaos.2020.109944.
26. Ozsahin, I., Sekeroglu, B., Musa, M.S., Mustapha, M.T., Ozsahin, D.U. Review on diagnosis of COVID-19 from chest CT images using artificial intelligence. *Computational and Mathematical Methods in Medicine* vol. 2020 (2020): ArticleID 9756518. doi: 10.1155/2020/9756518.

6 Machine Learning Approaches for Analysis in Healthcare Informatics

Jagadeesh K. and A. Rajendran

CONTENTS

6.1 INTRODUCTION

Machine learning (ML) is a technique that is mainly used for information analysis, and that can be mechanized as part of a system, resulting in a kind of man-made problem-solving technique, with negligible human intervention. As a result, new techniques have evolved, with ML acting as a division of artificial intelligence (AI) nowadays. One of the greatest transformative advances and empowering influences for human culture of this century will prove to be AI. It is agreed that AI and related technologies are set to change life for the better, improving ways of life. It is no mystery that this change is being achieved, to a great extent, by ML apparatus and strategies. ML, then, is about making computers modify or adapt actions so that predictions become more accurate.

6.1.1 Learning

Learning is the main thing that gives us adaptability in our life, regardless of how old we are. The significant elements of learning are recollecting, adjusting, and summing up. The final word, "summing up", is tied in with recognizing similitude in different circumstances, so that things used in one place can be used in another. This all makes learning helpful, in light of the fact that we can employ our insight in many different situations. There are plenty of different aspects, for example, of thinking and coherent finding, which we will not stress a lot in this chapter. We will concentrate on the major pieces of insight learning and adjustment, and how we can demonstrate them on a PC. There has additionally been considerable enthusiasm for making PCs reason and find solutions to problems. This was the premise of most early AI and is sometimes referred to as initiative handling on the grounds that the PC controls images that mirror nature. Interestingly, AI strategies are sometimes said to be sub-emblematic, in light of the fact that no images or representative control are included [1].

6.2 MACHINE LEARNING

Consider a situation where a model expects that a machine needs to predict whether a client will purchase a particular item, let's say "Antivirus", this year or not. The machine will do this by examining past information/ experiences, i.e., the information of items that the client had purchased each year and, in the event that he purchases Antivirus consistently, at that point, there is a high likelihood that the client is going to purchase Antivirus (or antivirus software) this year too. This is the means by which AI works at the fundamental calculated level. AI is fundamentally making human operations increasingly predictable. AI can identify truly life-changing potential in human operations, particularly in the area of clinical systems. AI involves getting PCs to program themselves. In the case that writing computer programs is computerization, at that point AI is robotizing the procedure of robotization. Composing programs is the bottleneck; we need more great program designers. Let the information accomplish the work, rather than individuals. AI is the best approach to making programming adaptable [2].

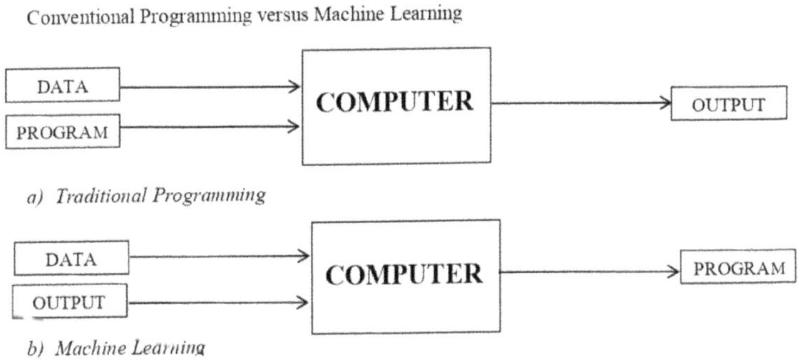

FIGURE 6.1 Conventional programming *versus* machine learning.

Conventional Programming: Data and program are run on the PC to create the yield [3]. **AI:** Data and its yield are run on the PC to make a program. This program can be utilized in conventional programming as shown in Figure 6.1. AI resembles cultivation or planting.

Uses of Machine Learning

Test uses of AI

- **Web search:** Positioning the page dependent on what you are destined to tap on.
- **Computational science:** Same-structure drugs in the PC dependent.
- **Accounts:** Conclude to whom which charge card offers should be sent. Step-by-step instructions should be presented to choose where to lodge cash.
- **Web-based business:** Regardless of whether an exchange is false.
- **Space investigation:** Space tests and radio stargazing.
- **Apply autonomy:** How to deal with vulnerability under new conditions. Self-governing. Self-driving vehicles.
- **Data extraction:** Ask inquiries over databases over the web.
- **Interpersonal organizations:** Data on connections and inclinations. Use of AI to extricate an incentive from information.
- **Investigating:** Corrected sentence: Use in software engineering issues like troubleshooting, work serious procedure, could recommend where the bug could be [4].

6.2.1 Types of Machine Learning

ML has been classified into four types [5], which are applicable in various fields, namely:

- **Supervised learning:** Task driven (regression/classification).
- **Unsupervised learning:** Data driven (clustering).

- **Reinforced learning:** Algorithm learns to react to an environment.
- **Evolutionary learning:** Biological evolution.

6.2.2 Different Algorithms

There are three algorithms, namely:

- Supervised
- Unsupervised
- Reinforced

6.3 SUPERVISED LEARNING

As mentioned earlier, AI accepts information as input, and these data are referred to as training data. The preparation data incorporates two items, inputs and labels (results). What are inputs and labels? The training model, with a cluster of preparation data (inputs and targets), has new information at that point and the rationale we reached before we predicted the yield [6]. We do not find 6 as solution we might get esteem which is near 6 dependent on preparing data and calculation [7]. The supervised learning concept and various types are shown in Figures 6.2 and 6.3.

6.3.1 Regression

The same issue arises in predicted results from a relationship.

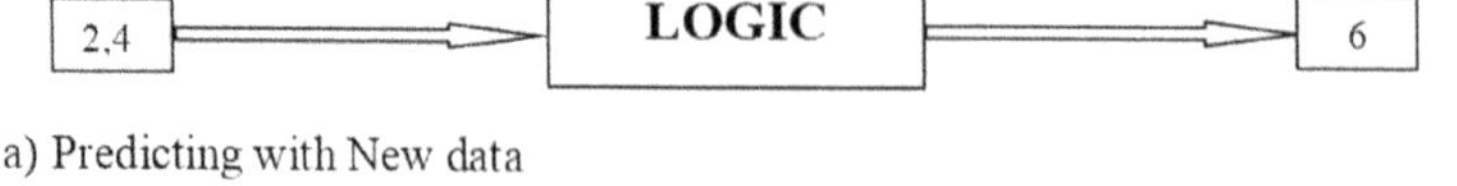

FIGURE 6.2 Supervised learning concept.

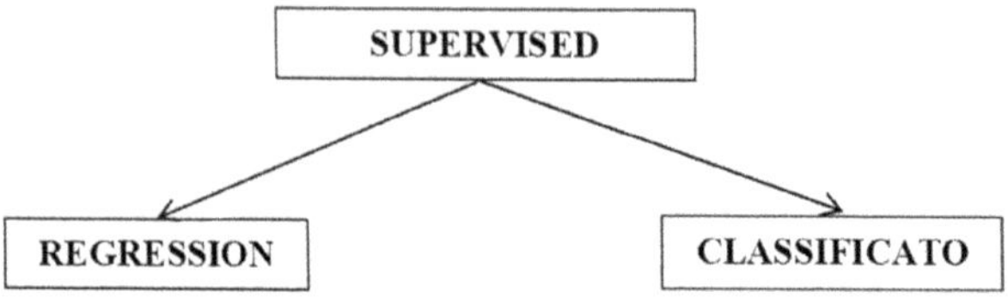

FIGURE 6.3 Types of supervised learning.

A few models of regression are

- What is the cost of living in a particular place?
- What is the estimation of expenditure?
- What is the total number of runs in a cricket match?

6.3.2 Classification

Arrangement: This is the sort of issue where we anticipate a clear-cut reaction, where the information can be grouped into explicit "classes" (e.g., we predict one of the qualities from a number of qualities) [4]. Regression and classification are depicted in Figure 6.4.

Essentially "True/False" types of situations are known as "paired arrangements".

6.4 UNSUPERVISED LEARNING

In unsupervised learning, the preparation information excludes results that advise the framework. This preparation information is not organized (e.g., it contains unexpected data, unknown information, etc.). The unsupervised process is shown in Figure 6.5.

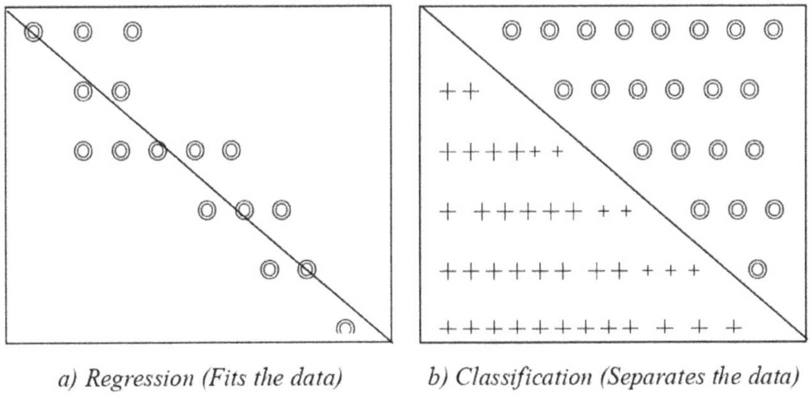

FIGURE 6.4 Regression and classification.

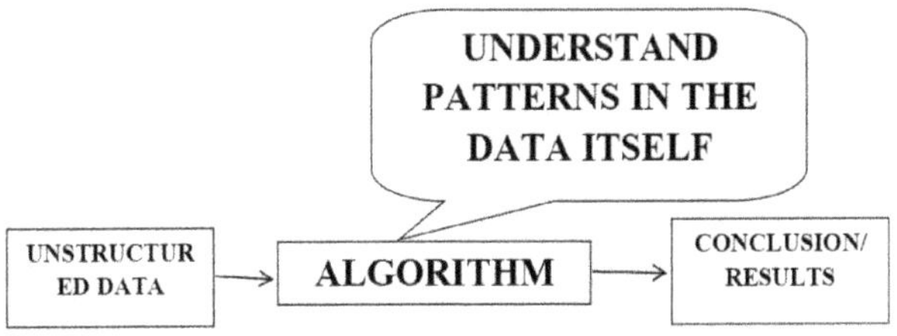

FIGURE 6.5 Unsupervised learning process.

Ex: An irregular article from various pages

Similarly, there are various sorts of clusters and inconsistency recognition factors (clusters are well known)

Clusters:

This is a situation where comparable things are grouped together. Like multi-class orders, however, here we do not give marks, but the framework understands information [6, 7].

Estimated number of clusters: 3

Unsupervised learning is a little hard to actualize, but it is shown in Figure 6.6.

Clustering with three clusters is depicted in Figure 6.7.

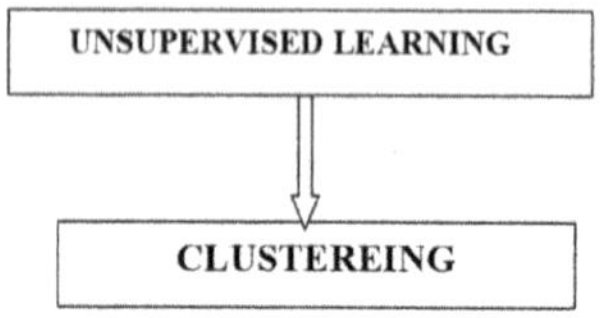

FIGURE 6.6 Unsupervised learning type.

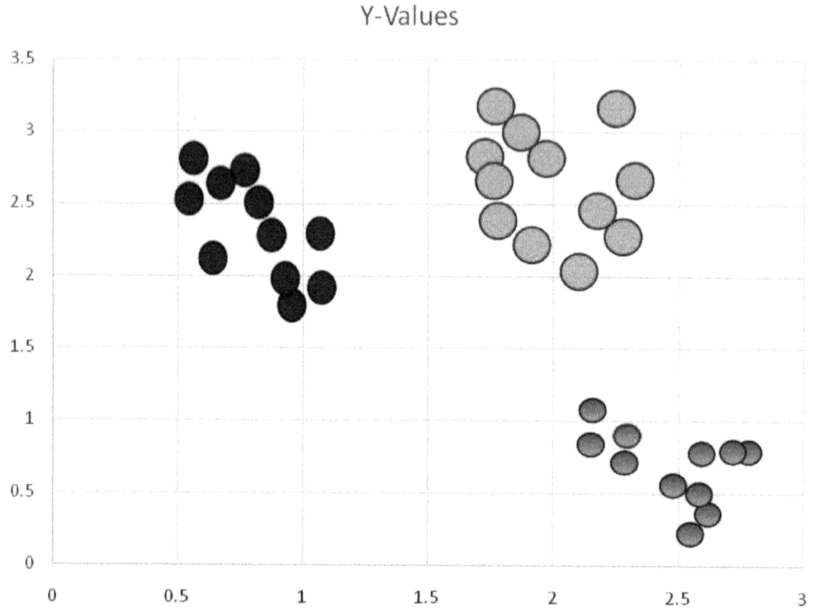

FIGURE 6.7 Clustering with three clusters.

6.4.1 *K*-Means Algorithm

K-means clustering is a type of unsupervised learning; where clustering is carried out, the algorithm finds a fixed number (*K*) of groups in a large volume of data. The cluster is a collection of data focusing on the individual groups [12].

Points of interest of *K*-means: This is a widely used straightforward strategy for group examination, which can be carried out rapidly.

6.4.2 Self-Organizing Feature Map (SOM)

Self-organizing maps or Kohenin's guide is a sort of fake neural systems presented by Teuvo Kohonen during the 1980s. We employ this type of fake neural system in measurements, that decrease to reduce our information, by making a spatially sorted out portrayal [13] (Figure 6.8).

Self-sorting out guide preparation:

SOM does not use back-propagation with SGD to refresh weightings; this kind of artificial neural network system utilizes competitive analysis to refresh its weightings. Competitive learning depends on the following three procedures:

- Competition
- Collaboration
- Adjustment

6.5 EVOLUTIONARY LEARNING

6.5.1 Genetic Algorithm

A genetic algorithm (GA) replicates the procedure for normal choice in a crowd to reach the exact person without any error or by reaching the wrong person. GAs are unique heuristic hunt algorithms. GAs depend on characteristic genetic qualities. Evolutionary learning is depicted in Figure 6.9.

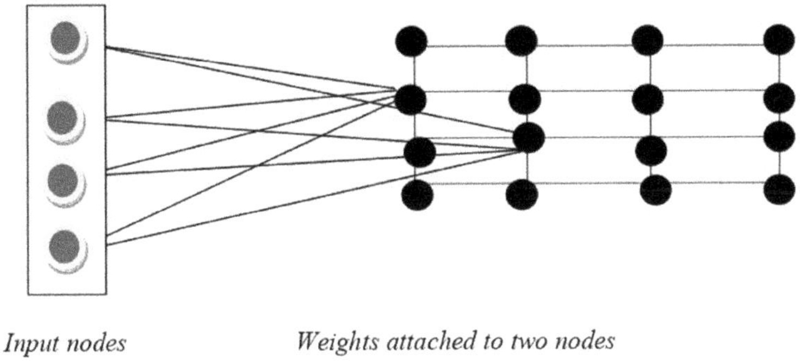

FIGURE 6.8 Self-organizing map network.

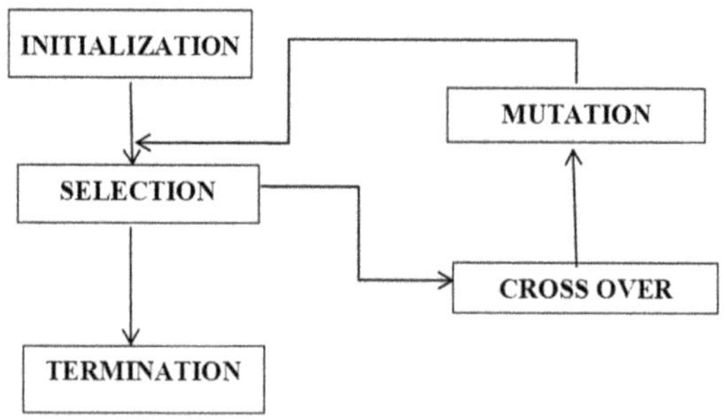

FIGURE 6.9 Evolutionary learning.

6.5.1.1 Establishing the Genetic Algorithm

The GAs were based on the structure of genetic models with chromosome components. The establishment of the GA-dependent relationship occurs as follows:

- Regions are individually populated.
- People with the most positive approach live for a long time, compared with others.
- Qualities that are mainly based on the "health"-related concept are very important and show a vast difference.
- Consequently, all the average calculations are based on the age of a particular person.

The numbers of inhabitants are maintained inside the pursuit space. Every individual will look for a space and answer the issues [14]. The GA is depicted in Figure 6.10.

6.6 REINFORCEMENT LEARNING

Reinforcement learning was characterized as a ML strategy concerned with how programming operators should take action in a domain. Fortification learning is an aspect piece of the learning strategy that causes amplification of part of the system. This neural network learning strategy encourages you to determine how to accomplish a goal or augment a particular measurement over numerous means. Reinforcement learning is shown in Figure 6.11.

Here are some significant terms employed in reinforcement learning:

Operator: It is an accepted element which performs activities in a situation to achieve some reward.

Condition (e): A situation that an operator needs to confront.

Prize (R): A quick return given to a specialist when the person performs an explicit activity or assignment.

State (s): "State" alludes to the current circumstance returned to the Earth.

Strategy (π): It is a system used for reinforcement in AI.

Worth (V): It is normal long-term, which comes back with rebate and contrasted with the momentary prize.

Worth Function: It indicates the estimation of an expression, that is the aggregate sum of the remuneration.

Model of the Earth: This mirrors the conduct of the Earth. It encourages you to make inductions and furthermore to decide how the Earth will carry on.

Model-based strategies: It is a technique for taking care of fortification learning issues which utilize model-based techniques.

Q worth or activity esteem (Q): Q esteem is very like "worth". The main distinction between the two is that Q accepts an extra boundary as a current activity.

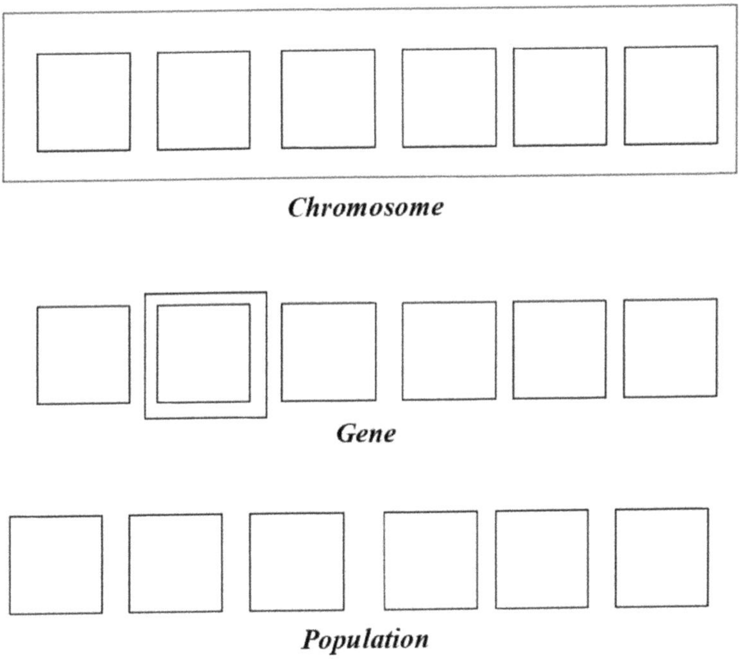

FIGURE 6.10 Genetic algorithm.

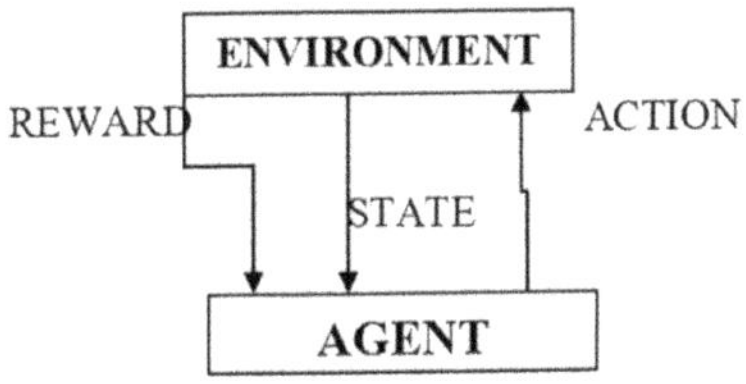

FIGURE 6.11 Reinforcement learning.

States: S

Model: $T(S, a, S') \sim P(S' \mid S, a)$

Actions: $A(S)$, A

Reward: $R(S)$, $R(S, a)$, $R(S, a, S')$

Policy: $\Pi(S) \longrightarrow a$

Π^*

FIGURE 6.12 Markov decision process.

6.6.1 Markov Decision Process

Fortification learning is a type of ML [12]. The model contains the following parameters and it is shown in Figure 6.12.

- Similar states are represented as "S"
- Models are in the form of sets
- Actions that have possibility are represented as "A"
- The true value is represented as "R(s, a)"
- Final result of the process

6.7 HEALTHCARE INFORMATICS

6.7.1 Health Care

Healthcare informatics create a collection of information concerned with the use of data that correspond to innovation inside clinical exploration and instruction, and, in addition, for advancing medical services. The field centers on the biomedical data. Obtaining the capacity to recover the ideal use of data for critical thinking is dynamic [11].

The objective of healthcare informatics is to improve health in conjunction with caring specialists, improving the methods of operation and analysis. Different forms of use of clinical informatics include telemedicine, E-medical records, information retrieval, and decision making as the most important techniques. Various uses of clinical informatics will achieve the best results to consider in terms of treatment, will reduce expenses in social insurance administrations with a decrease in error, and will provide the data needed by the doctors, including updated data and information.

6.7.2 Applications

The steady increase in the development of the number of uses of AI in social insurance permits the examination, further details, developments, and previous medical history of a person without a regular update. Later, the ML concepts were adapted to normal procedures and the frameworks can operate in various fields [23].

The ten best uses of ML in medicine:

1. **Distinguishing Diseases and Diagnosis**
   The main difference between the disease and diagnosis deals with the delay in curing a particular patient. The disease may be critical and the entire process will be clearly understood with the help of diagnosis.
2. **Medication Discovery and Manufacturing**
   Medical discoveries include recent as well as historical developments They are capable of curing a specific disease in a short time. With all these innovative concepts, it becomes easier to develop a new pharmaceutical chemical and formulate it as a drug.
3. **Clinical Imaging**
   AI and deep learning are amenable to the advanced technique known as computer vision. They complement the inner-eye process by dissecting images to achieve more detailed analysis. As the informative limit of AI increases, it is hoped that data from different clinical notations can be included *via* AI-driven indicative procedures.
4. **Customized Drugs**
   Customized drugs (as part of personalized medicine) will improve the health benefits for the patient. Nowadays, doctors can identify diseases at an early stage. There may be some hazards and hurdles based on the patient's past history, needing regular supervision and further treatment of the patient.
5. **AI-Based Behavioral Modification**
   Social differences are a significant part of preventive drugs. In AI-based medical services, new businesses are continually emerging in the fields of disease, prediction, diagnosis, tolerance of medical treatment, and so on.
6. **Medical Records**
   Storing and being able to retrieve important medical documents is a comprehensive procedure. A new innovation concept has been introduced that processes the data so that they can be handled by ML.
7. **Clinical Research and Trials**
   ML-related investigations to predict effective medical treatments can assist analysts by drawing from a wide assortment of information focuses, for example, past specialist visits, online networking, and so forth. AI has likewise developed a use in guaranteeing constant checking and information access, identifying the best sample size, and utilizing electronic health records to decrease information-based mistakes.
8. **Publicly Supported Data Collection:**
   Data collection and storage is a very tedious process, but, with the help of ML, it can become very simple, effective, and user-friendly too. The public details must be stored and maintained with care, especially aadhar details,

bank account details, and other personal information. Once data are stored, they must be maintained securely and must be retrieved or deleted, whenever necessary.

9. **Best Radiotherapy**
Clinical image analysis has numerous factors, providing a specific snapshot in time. The numerous sores, symptoms, and so on that cannot be demonstrated involve problematic conditions. Although ML-dependent calculations result in an increase in the volume of accessible data, ML makes it simpler to analyze and interpret the facts. One of the best-known uses of AI in clinical image examination is the characterization of items, for example, injuries, into classes, such as typical or anomalous, injured or non-injured, and so forth.
10. **Episode Identification**
Episode identification includes a summary of the steps incuded in ML concepts, providing the clear steps necessary to be followed in identification. The process explains the step-by-step activities that take part in the operation.

6.8 ANALYSIS AND DIAGNOSIS

Providing the correct analysis is a key part of human services. Errors will affect the patient by deferring the appropriate treatment as well as affecting the medical records negatively, which, in turn, could cause finance-related issues [31, 32].

6.8.1 Analysis

The healthcare investigation of complex frameworks involves collaborations between different types of specialists. The following steps of the investigation need to be directed [25].

(1) Specification of the types of specialists which act in the framework, with regards to their functionalities.
(2) Specifically, what kind of framework, as indicated by Mazur's hypothesis [27], relates to a given operator, i.e., whether it is a self-ruling framework, or rather completely or halfway controlled, and how the specialist's inward modules (correlator and gatherer) are composed.
(3) Analysis of the points concerning the individual types of operators, both the strategic (present moment) points and the vital (long-term) ones.
(4) The particulars identified can be introduced as a graph stream algorithm, which is a typical technique employed in computer science.
(5) Analysis of the relations and communications described previously examines the vitality of scattering in singular stream and postponement of streams, admissions and control activities.
(6) Analysis of the entire framework, including

(a) The type of framework, as indicated by Mazur's hypothesis, represented by the entire framework
(b) Analysis, regardless of whether there exist special operators in the framework
(c) Analysis, regardless of whether the entire framework is acceptable or not, i.e., whether individual specialists collaborate or compete

6.8.2 Diagnosis

An exact diagnosis implies that a specialist realizes how to make the correct decisions for their patients at the correct time. It implies that individuals realize they can have improved command over their own health. In addition, it implies that the human services frameworks realize that they are putting resources into the correct treatments, that have a genuine effect. At the point when an appropriate diagnosis is settled on, the correct treatment decisions can then be made. For instance, there are a few unique sorts of breast disease, each one of which requires an alternative treatment approach. A precise diagnosis opens up the route to the correct treatment.

Preferably, an analysis ought to be precise, yet also opportune. An early and precise diagnosis can essentially increase a patient's survival period. In malignant breast cancer, for instance, 98.8% of patients survive for at least five years whenr diagnosed early, contrasting with just 26.3% when diagnosed at a late stage. However, in some African nations, 80% of patients are diagnosed at the late-to-end stage. For certain diseases, including numerous sorts of cancers, diagnosis can be unpredictable, as it requires expert researchers or laboratories for testing [27].

Emergency clinics or centers in numerous regions of the world are just not equipped with the equipment or expert staff necessary for accurate diagnosis. With rarer ailments, a finding is regularly postponed or mistaken in light of the fact that side effects can be mistaken for those of other basic conditions. The absence of self-awareness is likewise a significant issue, since it can mean that patients do not notice the symptoms that would lead them to visit their primary care physician earlier. The healthcare analysis and its diagnosis process is depicted in Figure 6.13.

6.9 MACHINE LEARNING IN HEALTH CARE

6.9.1 Overview

The information received from the core and the data associated with the human operator initiates the learning stage. In any case, in spite of the fact that human operators can be excellent at dealing with up to three factors, the biomedical informational indexes involve data consisting of much more than three factors, making manual information analysis difficult. Fruitful use of AI in health informatics requires consideration of the entire pipeline from information preprocessing to information representation [12].

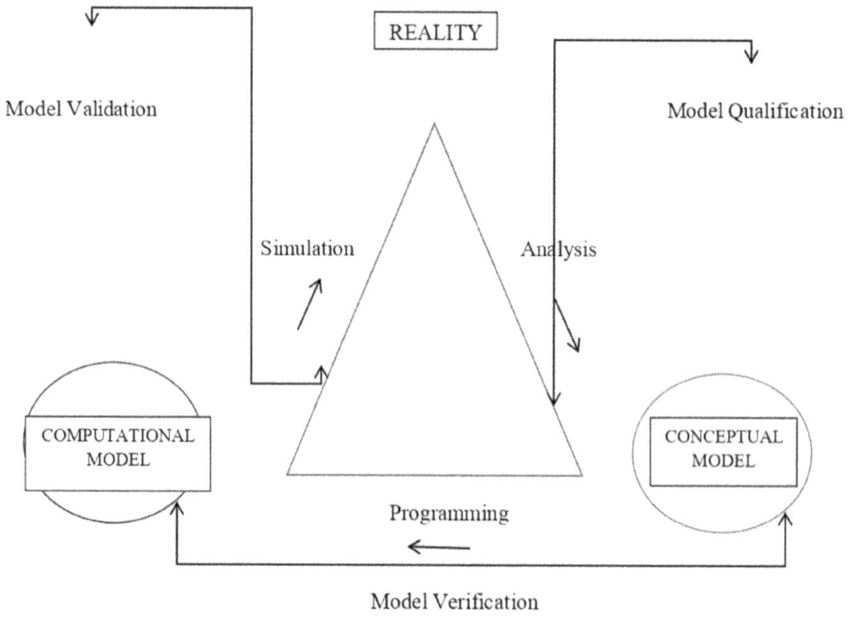

FIGURE 6.13 Healthcare analysis and diagnosis.

6.9.2 Types

Computerized Imaging

The productivity of laboratory testing, combined with human interpretation, is certainly higher, using computers that are loaded with programs for advanced imaging. With regard to diagnostics, human brainpower-based diagnostics, combined with AI-based clinical analysis, offers a shorter turnaround time than does conventional testing.

A Cloud-Based Platform for Interaction

Through AI-based clinical analysis, pathologists can send information and images to authorities and experts around the world, to upgrade analysis and diagnosis, without the need to physically move the sample or patient from one area to another. The highlights of AI in pathology, including AI for clinical testing, decrease the frequency of mistakes that are frequently seen in physical manual investigations.

6.9.3 Applications

The goal for ML is to make it increasingly effective, precise and rapid. In a social insurance framework, the ML system is the specialist's mind and information, since, in every case, a patient still needs a human touch and care. Neither ML nor some other innovation can supplant this. A robotized machine can offer the clinical and technical supports in a better way. Below, the ten main uses of ML in human services are listed [7]:

- Coronary disease diagnosis
- Prediction of diabetes
- Forecast of liver disease
- Robotized surgery
- Malignant growth detection and prediction
- Customized medical treatment
- Pharmaceutical discovery
- Storage and recovery of electronic health records
- AI in radiology
- Clinical trials and research

6.9.4 Modules

ML speed up the logical disclosure in situations where the diagnostic process is not a special case. Developments which speed up the examination of the patient, and the availability of prescient calculations, can prepare clinical staff for a looming cardiovascular failure [17]. ML supplements human knowledge and practice across clinical systems and is depicted in Figure 6.14.

Machine-learning modules:

1. Regular language processing
2. Understanding of risk identification
3. Accelerating medical research insights
4. Computer vision for cancer cell identification
5. Utilizing convolutional neural networks (CNN)

6.10 CONCLUSION

The benefits of ML have been widely discussed in this chapter. AI and ML together will play a vital role in future healthcare informatics, with new innovative techniques. Current conditions of human services do not provide the motivation to share information on the framework. A human services transformation is in progress in the USA, to revolutionize information processing and analysis. Comparable methodologies are being investigated in China. Man-made reasoning (ML) is changing the manner in which we associate and analyze data, and develop diagnostic and treatment planning products and enterprises across ventures. The eventual fate of human involvement in health services systems includes:

- A human services system arranged around the outline of man-made consciousness (AI), regular language processing (NLP), and AI (ML)
- ML-based social insurance may expand as these innovations affect the act of medical intervention and human services throughout the coming decade.

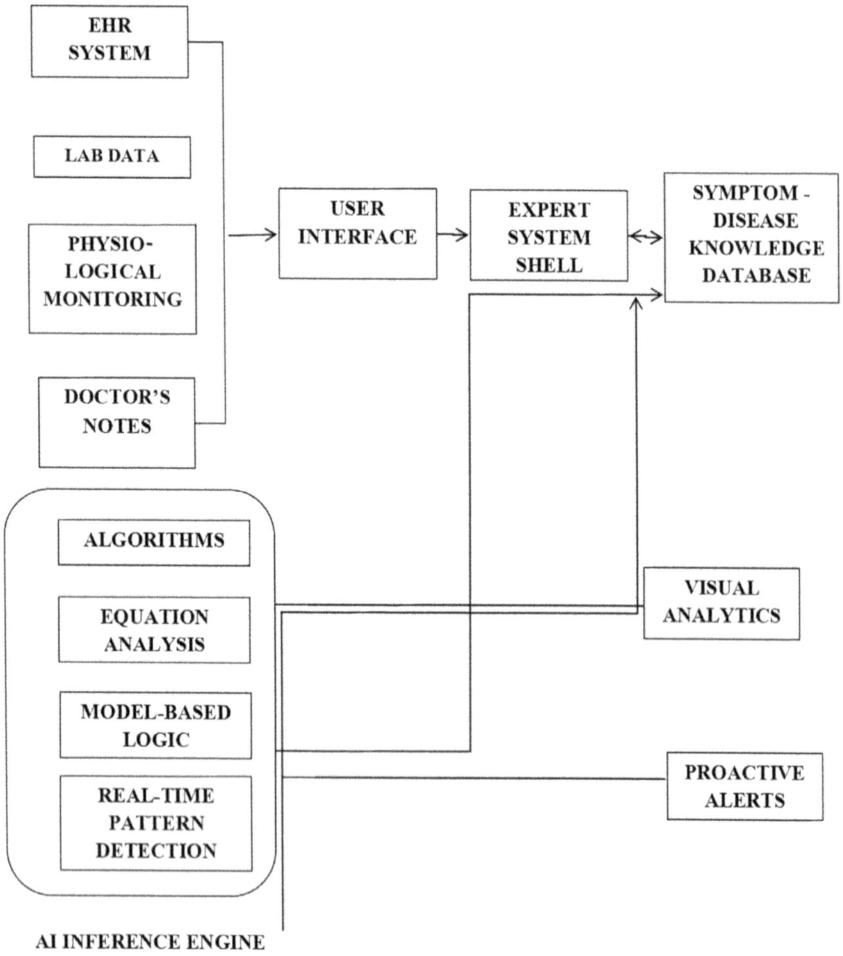

FIGURE 6.14 Machine learning module in health care.

REFERENCES

1. M. J. Abinash and V. Vasudevan, "A study on wrapper-based feature selection algorithm for leukemia dataset." In *Intelligent Engineering Informatics*. Springer, Singapore, 2018, pp. 311–321.
2. L. Chebouba, D. Boughaci, and C. Guziolowski, "Proteomics versus clinical data and stochastic local search based feature selection for acute myeloid leukemia patients' classification." *Journal of Medical Systems* vol. 42, no. 7, p. 129, 2018.
3. S. Lee, et al., "A novel bagging C4. 5 algorithm based on wrapper feature selection for supporting wise clinical decision making." *Journal of Biomedical Informatics* vol. 78, pp. 144–155, 2018.
4. S. Venkataraman and R. Selvaraj, "Optimal and novel hybrid feature selection framework for effective data classification." In *Advances in Systems, Control and Automation*. Springer, Singapore, 2018, pp. 499–514.

5. P. Chaudhari and H. Agarwal, "Improving feature selection using elite breeding QPSO on gene data set for cancer classification." In *Intelligent Engineering Informatics*. Springer, Singapore, 2018, pp. 209–219.
6. H. R. Roth, et al., "Improving computer-aided detection using convolutional neural networks and random view aggregation." *IEEE Transactions on Medical Imaging*, vol. 35, no. 5, pp. 1170–1181, 2016.
7. R. Fakoor, F. Ladhak, A. Nazi, and M. Huber, "Using deep learning to enhance cancer diagnosis and classification." In *Proc. Int. Conf. Mach. Learn.*, 2013, pp. 1–7.
8. B. Alipanahi, A. Delong, M. T. Weirauch, and B. J. Frey, "Predicting the sequence specificities of DNA-and RNA-binding proteins by deep learning." *Nature Biotechnology*, vol. 33, pp. 831–838, 2015.
9. Pinaki Pratim Acharjya, Santanu Santra, and Dibyendu Ghoshal, "An improved scheme on morphological image segmentation using the gradients." *(IJACSA) International Journal of Advanced Computer Science and Applications*, vol. 4, no. 2, 2013.
10. Y. LeCun, Y. Bengio, and G. Hinton, "Deep learning." *Nature*, vol. 521, no. 7553, pp. 436–444, 2015.
11. H. Wang, Q. Zhang, and J. Yuan, "Semantically enhanced medical information retrieval system: A tensor factorization based approach." *IEEE Access*, vol. 5, pp. 7584–7593, 2017.
12. M. N. Ahmed, A. S. Toor, K. O'Neil, and D. Friedland, "Cognitive computing and the future of health care cognitive computing and the future of healthcare: The cognitive power of ibm watson has the potential to transform global personalized medicine." *IEEE Pulse*, vol. 8, no. 3, pp. 4–9, May 2017.
13. M. Abadi, A. Chu, I. Goodfellow, H. B. McMahan, I. Mironov, K. Talwar, et al., "Deep learning with differential privacy," In *Proceedings of the 2016 ACM SIGSAC Conference on Computer and Communications Security*, pp. 308–318, 2016.
14. S. Albarqouni, C. Baur, F. Achilles, V. Belagiannis, S. Demirci and N. Navab, "Aggnet: Deep learning from crowds for mitosis detection in breast cancer histology images," *IEEE Transactions on Medical Imaging*, vol. 35, no. 5, pp. 1313–1321, 2016.
15. Alzheimer's Association, "2018 Alzheimer's disease facts and figures." *Alzheimer's Dementia*, vol. 14, no. 3, pp. 367–429, 2018.
16. M. F. Mendez, "Early-onset Alzheimer disease." *Neurologic Clinics*, vol. 35, pp. 263–281, 2017.
17. K. A. Jellinger, "Dementia with lewy bodies and Parkinson's disease-dementia: Current concepts and controversies." *Journal of Neural Transmission*, vol. 125, no. 4, pp. 615–650, 2018.
18. M. Liu, D. Zhang and D. Shen, "Relationship induced multi-template learning for diagnosis of Alzheimer's disease and mild cognitive impairment.", *IEEE Transactions on Medical Imaging*, vol. 35, no. 6, pp. 1463–1474, June 2016.
19. A. Arab, A. Wojna-Pelczar, A. Khairnar, N. Szab and J. Ruda-Kucerova, "Principles of diffusion kurtosis imaging and its role in early diagnosis of neurodegenerative disorders." *Brain Research Bulletin*, vol. 139, pp. 91–98, 2018.
20. P. Bennett and N. R. Hardiker, "The use of computerized clinical decision support systems in emergency care: A substantive review of the literature." *Journal of the American Medical Informatics Association* vol. 24, no. 03, pp. 655–68, 2017.
21. A. Wright, A. Ai, J. S. Ash, J. Wiesen, H. T. Hickman, S. Aaron et al., "Clinical decision support alert malfunctions: Analysis and empirically derived taxonomy." *Journal of the American Medical Informatics Association* vol. 25, no. 05, pp. 496–506, 2018.
22. A. Goldstein, Y. Shahar, E. Orenbuch, M. J. Cohen, "Evaluation of an automated knowledge-based textual summarization system for longitudinal clinical data, in the intensive care domain." *Artificial Intelligence in Medicine* vol. 82, pp. 20–33, 2017.

23. H. R. Roth et al., "Improving computer-aided detection using convolutional neural networks and random view aggregation." *IEEE Transactions on Medical Imaging* vol. 35, no. 5, pp. 1170–1181, May 2016.
24. B. Ramsundar, S. Kearnes, P. Riley, D. Webster, D. Konerding and V. Pande, "Massively multitask networks for drug discovery." Feb. 2015.
25. S. Zhang et al., "A deep learning framework for modeling structural features of rna-binding protein targets." *Nucleic Acids Research* vol. 44, no. 4, pp. e32–e32, 2016.
26. K. Tian, M. Shao, S. Zhou and J. Guan, "Boosting compound-protein interaction prediction by deep learning." In *Proceedings of IEEE International Conference on Bioinformatics*, pp. 29–34, 2015.
27. C. Angermueller, H. Lee, W. Reik and O. Stegle, "Accurate prediction of single-cell DNA methylation states using deep learning. " bioRxiv, 2016.
28. J. Shan and L. Li, "A deep learning method for microaneurysm detection in fundus images." In *Proc. IEEE Connected Health Appl. Syst. Eng. Technol.*, pp. 357–358, 2016.
29. A. Mansoor et al., "Deep learning guided partitioned shape model for anterior visual pathway segmentation." *IEEE Transactions on Medical Imaging* vol. 35, no. 8, pp. 1856–1865, Aug. 2016.
30. D. Nie, H. Zhang, E. Adeli, L. Liu and D. Shen, "3d deep learning for multi-modal imaging-guided survival time prediction of brain tumor patients." In *Proc. MICCAI*, pp. 212–220, 2016.
31. J. Kleesiek et al., "Deep MRI brain extraction: A 3D convolutional neural network for skull stripping." *NeuroImage* vol. 129, pp. 460–469, 2016.
32. K. Fritscher, P. Raudaschl, P. Zaffino, M. F. Spadea, G. C. Sharp and R. Schubert, "Deep neural networks for fast segmentation of 3d medical images." In *Proc. MICCAI*, pp. 158–165, 2016.

7 Prediction of Epidemic Disease Outbreaks, Using Machine Learning

Vaishali Gupta and Sanjeev Kumar Prasad

CONTENTS

7.1 INTRODUCTION

An epidemic is the spread of a disease more rapid than would be expected in a community or region over a given time period. Epidemic diseases are contagious and have the capacity to affect an entire nation if they are not managed before reaching their outbreak level. Such diseases can endanger the lives of the whole population. There have been a number of epidemic outbreaks occurring in various countries in recent years. Some of these are well-known outbreaks, including dengue, cholera, diphtheria, influenza, and bird flu [1–8]. Health data suggest that contagious diseases are responsible for 43% of deaths worldwide [7]. India has been affected by epidemic outbreaks and has witnessed outbreaks, such as dengue, chikungunya, swine flu, etc., in its various states and cities.

The recent outbreak of the pandemic coronavirus disease (COVID-19) is a clear example of the potential of epidemic outbreaks to derail not only a nation but also the entire world, with significant implications for lives and livelihoods. COVID-19, which started its spread in China initially in December 2019, had spread to almost the entire world within a few months. The virus is spreading rapidly, despite lockdown measures and other efforts to control and contain the virus.

Epidemic diseases have significant impacts on communities' health across the world and hence there is an increasing pressure for predicting, as well as preventing, such diseases. In such situations, prediction can serve as the starting point in decision-making processes for the response to an epidemic disease outbreak. Prediction of disease outbreaks provides warning that a certain amount of disease may occur at a particular time in the future. Prediction of disease occurrence ensures that control measures are used more effectively.

Before the advent of computer-driven data-intensive analytical techniques, predictions were achieved using classical models of data analysis. However, the growing use of computer-aided techniques in predictive analysis has helped in not only finding existing patterns in the data but also revealing their future behavioral trends. These techniques are based on the acquisition of huge volumes of relevant data, and require computers, software, and algorithms to learn and train through the use of more and more of such data, to increase the accuracy of future predictions, a strategy known as machine learning techniques.

Due to the availability of huge amounts of digital data in healthcare, machine learning techniques have emerged as popular choices in healthcare epidemiology and are increasingly used, not only for predicting and detecting disease trends and patterns, but also for identifying disease risk factors in the population. Digital healthcare epidemiology uses digitally generated data to understand the distribution pattern and determining factors for health and disease conditions in designated populations.

Against this backdrop, this chapter discusses the application of machine learning techniques in the healthcare sector for the prediction of epidemic disease outbreaks. The chapter is structured as follow. After the introduction, Subsection 2 discusses predictive analytics and its role in the healthcare industry. Subsection 3 explains the concept of machine learning and highlights the process involved and various types of machine learning algorithms. Subsection 4 discusses machine learning models in the context of predicting the outbreak of epidemic diseases, whereas Subsection 5 highlights various dissemination factors for an epidemic outbreak, and Subsection 6 discusses machine learning algorithms that have been commonly applied in the literature for predicting such outbreaks. Subsection 7 provides a glimpse of the existing research on machine learning applications to epidemic prediction, and Subsection 8 highlights the challenges and opportunities facing the use of machine learning techniques in real-life cases of disease outbreaks. Finally, Subsection 9 concludes the chapter by highlighting the potential use of machine learning techniques in containing the current outbreak of COVID-19.

7.2 PREDICTIVE ANALYTICS

Predictive analytics is a data analytics process that aims to make predictions about a future outcome, based on historical data and analytical techniques, such as statistical analysis and machine learning. Predictive analytic modeling is used to find and exploit patterns contained in the historical data in order to identify current risks and future opportunities and challenges. Thus, predictive analytics is a forward-looking process that can be used to predict future events based on what has happened in the past and what is happening at present [9]. Figure 7.1 provides a snapshot of the predictive analytics process.

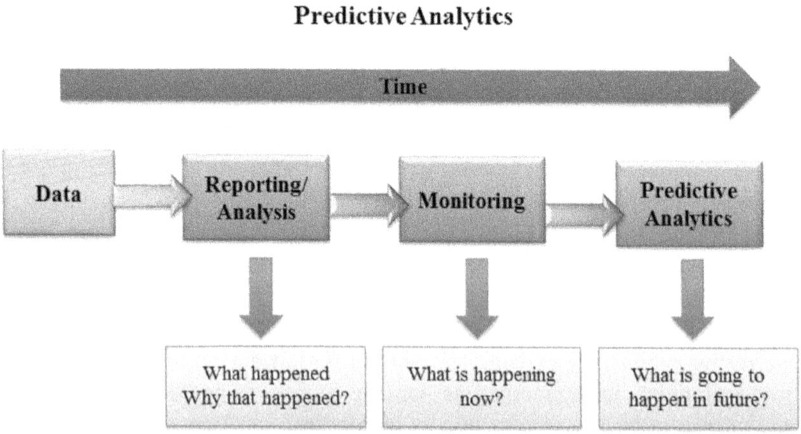

FIGURE 7.1 Predictive analytics process.

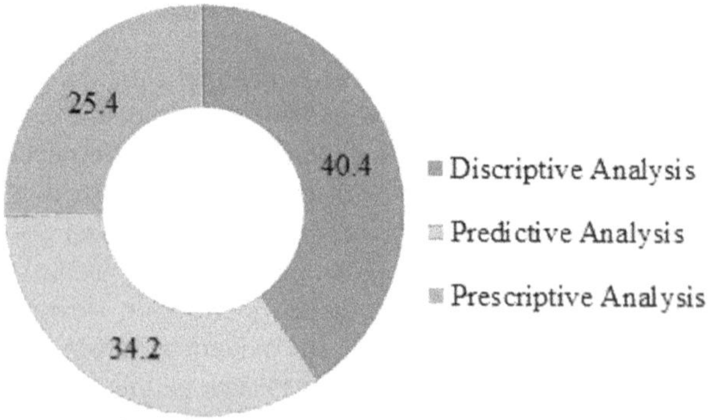

FIGURE 7.2 Global healthcare big data analytics market.

7.2.1 Role of Predictive Analytics in Healthcare

Predictive analytics can be used in the healthcare industry to help health practitioners to transform the huge volume of health-related data into actionable insights that can improve decision making in health management. Nowadays, most health organizations are developing big data analytics capabilities, such as machine learning algorithms, to extract actionable information from growing datasets to forecast future events in real time in order to improve public health, as well as the health of individuals.

Figure 7.2 shows the use of big data analytics in the global healthcare industry. Prescriptive analytics accounts for 40.4%, predictive analytics for 34.2%, and descriptive analytics for 25.4% of the global healthcare market. Thus, predictive analytics plays a significant role in the healthcare industry [10].

Predictive analytics benefits different segments of the healthcare industry, such as public health management, healthcare providers, health insurance providers, and the individual patients and customers. In the case of public health, predictive analytics enables health providers to identify those individuals with elevated risks of infectious disease and helps them to create risk scores, based on patient data, laboratory tests results, claimed data, and clinical data for future epidemic disease outbreaks.

Predictive analytics enables the epidemic control centers to track public health in real time, watching for any emerging disease trends and patterns that could lead to epidemic disease outbreaks. Thus, disease control and management are also application areas where predictive analytics helps environmental health practitioners through providing early warning systems for any suspicious symptoms among the population.

7.3 MACHINE LEARNING

Machine learning is a tool for transforming data into information. It is an application of artificial intelligence, which enables the machine to learn automatically from the data and improve itself from experience without being explicitly programmed to do so. Tom Mitchell described machine learning: "A computer program is said to learn from experience E with respect to some class of task T and performance measure P, if its performance at task T, as measured by P, improves with experience E" [11].

The main objective of machine learning algorithms is to construct a model from sample data for predicting future events, to improve a complex decision-making process.

7.3.1 Machine Learning Process

The machine learning process aims to predict future outcomes by optimizing the decisions in all kinds of decision making, based on the predictive information from large-scale data sets. Figure 7.3 represents the main steps, which will be described below, involved in a typical machine learning process.

7.3.1.1 Main Steps in Machine Learning Process

- **Gathering data:** Data collection plays a major role in the machine learning process because data are the foundation of any machine learning process, and the quality and quantity of gathered data directly determine how good a predictive model can be. Obtaining appropriate data depends on the problem in hand and the data type.

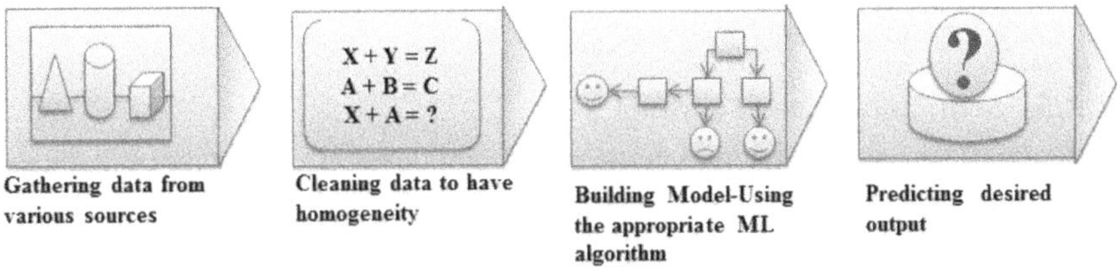

FIGURE 7.3 A typical machine learning process.

- **Cleaning data:** Real-world data often has unorganized, missing, or noisy elements. Therefore, data need to be cleaned, prepared, and organized in order to develop a successful prediction model through the machine learning process.
- **Model building:** This step involves selection of an appropriate machine learning algorithm to build a predictive model. This is an iterative step and the procedure will build several candidate models until a model good enough to be deployed for its intended task is developed.
- **Prediction:** Prediction or inference is the step where the model is ready to make predictions for the intended problem. This is the point at which the value of machine learning is realized.

7.3.2 Types of Machine Learning Algorithms

The American computer scientist Arthur Samuel defined 'machine learning' as "a computer's ability to learn without being explicitly programmed" [12]. The term "learning" in machine learning does not involve consciousness but learning in terms of recognizing different patterns or statistical regularities in the given large-scale data. Machine learning primarily aims to design algorithms that allow a machine to learn. These algorithms can be of four types, namely supervised, unsupervised, semi-supervised, or reinforcement learning.

7.3.2.1 Supervised Learning

These algorithms are designed to analyze the training data and to deduce an inferred function to map the new examples. The training data consists of training examples, with each example being a pair of input object and desired output values. In the supervised learning technique, machines are trained with well-labeled datasets (training datasets). Examples include logistic and linear regression, decision trees, support vector machines, naïve Bayes, etc. Supervised learning algorithms are the standard algorithms in epidemiological practice.

7.3.2.2 Unsupervised Learning

These algorithms are designed to identify inherent relationships and groupings inside the datasets without making any reference to either the output value or the "right answer" [13]. In unsupervised learning techniques, machines are trained with unlabeled datasets in order to allow algorithms to act upon those unlabeled datasets without any supervision. *K*-means clustering and association rules are some examples of this type of learning.

7.3.2.3 Semi-Supervised Learning

These algorithms are designed to make use of both labeled and unlabeled data, and machines are also trained with a combination of these two types of data. The cost associated with labeled data is high whereas obtaining unlabeled data is relatively inexpensive. By using both kinds of data, semi-supervised learning techniques simplify the problem associated with large labeled datasets. Examples of semi-supervised learning are speech analysis and internet content classifiers.

7.3.2.4 Reinforcement Learning

These algorithms are used for training machines to make specific predictions. The machine is exposed to the environment, with which it interacts by performing actions, and the machine keeps training itself through a trial-and-error approach. The machine continuously learns from previous experiences and attempts to take the best value for predicting future scenarios accurately. The Markov decision process is an example of a reinforcement learning algorithm.

7.4 MACHINE LEARNING MODELS FOR PREDICTING AN EPIDEMIC DISEASE OUTBREAK

As highlighted in Section 7.2.1, prediction is a statistical and analytical tool that has been increasingly used in recent disease outbreaks to help policy makers with decision making in real-time response to an outbreak, using machine learning algorithms that employ analytic data-based techniques for making predictions. Various steps in the machine learning process are mentioned in Subsection 7.3.1. In accordance with this, the prediction analysis of epidemic disease outbreak starts with gathering data from various data sources through a data mining process and goes through the data-based machine learning process to discover information in order to make predictions. Generation of epidemic disease outbreak predictions results follows a prediction workflow process. Figure 7.4 shows the machine learning process to predict epidemic disease outbreak, using epidemic disease data [14].

The prediction process for an epidemic disease outbreak, using machine learning techniques, encompasses the following three broad stages.

7.4.1 Collection and Cleaning of Epidemic Disease Outbreak Data

7.4.1.1 Data Collection

Data collection is an important part for any predictive analysis. Gathering up-to-date data is crucial for predictive analysis and efficient management of the outbreak. The data quantity, quality, and timeliness are important factors for epidemic disease prediction. Therefore, data from various sources need to be analyzed to make

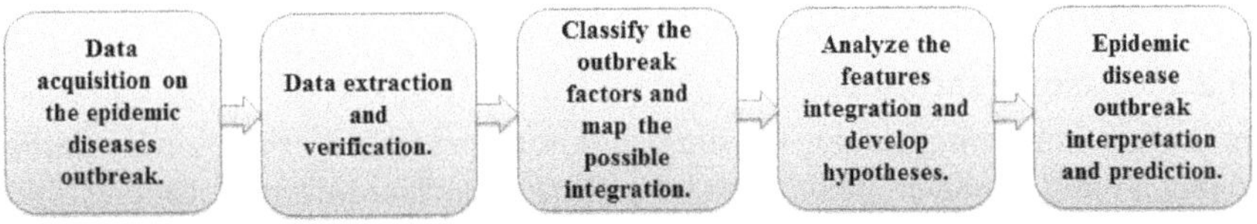

FIGURE 7.4 A machine learning process for epidemic disease prediction.

FIGURE 7.5 Data sources for epidemic disease outbreak.

predictions. Some of the popular data sources for epidemic disease outbreak include climatic data, demographic data, medical records, internet search queries, and social media data, such as Twitter. These data sources are described in Figure 7.5.

7.4.1.2 Data Cleaning

Data cleaning is a process of removal of unwanted observations, fixing structural errors, handling missing data, and managing unwanted outliers. Since data are collected from various sources, they are generally not ready for immediate analysis to support decision making. Hence, data must firstly be preprocessed and cleaned. For this purpose, technologies that convert unstructured, unprocessed, and raw data into structured forms are used. The cleaning of epidemic disease data includes extraction of data from clinical case studies or clinical notes in electronic health records, and the translation of data stored in non-standard formats into machine-readable standard formats.

7.4.2 Training the Model and Making Predictions, Using Machine Learning Predictive Analytics

7.4.2.1 Training the Model

There are a number of machine learning algorithms which have successfully trained the models and generated predictions for pathogens such as influenza, dengue, Zika, and Ebola diseases. The selection of algorithm to be used generally depends on the data type. It also takes into consideration the size and the dimensionality of the data.

7.4.2.2 Prediction

This is the stage where the trained model predicts the desired output, based on learning for a given input. It involves identifying a disease outbreak event corresponding to a possible outbreak in the community.

7.4.3 Results Visualization and Communication

Effective communication of prediction results ensures that they must produce actionable insights into the outcome. Visualizations also play an important role in the interpretation of prediction results in order to support decision making.

7.5 EPIDEMIC DISEASE DISSEMINATION FACTORS

The epidemic disease dissemination factors can be divided into four main categories: (i) the physical network, (ii) the geographical location, (iii) clinical studies, and (iv) social media. Figure 7.6 shows a pictorial representation of the epidemic disease dissemination factors that are to be considered in the prediction of an epidemic disease outbreak.

7.5.1 Physical Network

The spread of an infectious disease is mainly caused by harmful virus attacks. These viruses continuously search for host cells. Once they are inside the host cell, the viruses rapidly replicate and start to capture and hijack the ecosystem of the host's cell. Since the viruses are not capable of moving from one host cell to another on their own, they require a third medium to achieve such movement [6]. This mobilization of viruses from one host cell to another host cell is termed a physical network. Therefore, a physical network can be defined as the medium of virus transmission among host individuals. The main sources of the physical network are population density and hotspots.

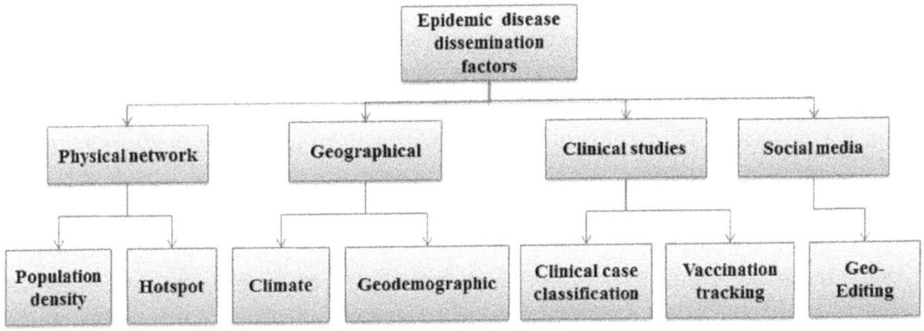

FIGURE 7.6 Disease epidemic dissemination factors.

7.5.1.1 Population Density

The density of the population is a measure of the number of persons living per unit area. The higher the value of this density in a specific area, the greater is the potential for epidemic disease outbreaks. Dense population areas are more prone to infectious disease transmission as close contacts are more likely to happen in such areas. The main factors which affect the health conditions of communities in populated areas are human migration, urbanization, and working conditions.

7.5.1.2 Hotspots

Hotspots are those areas where the probability of disease emergence is very high, such as tourist sites, amusement parks, places of historical significance, business centers, shopping malls, and airports. These hotspots play a key role in the rapid spread of contagious epidemic diseases because the possibility of closer contacts among the people is quite high in these areas, so that they become major risk factors in epidemic disease dissemination.

The transmission of viruses among hosts can be triggered by both direct and indirect contact of a healthy individual with a disease-infected person. The direct contact of infection includes human–human, human–animal, and animal–human contacts. The spread of a virus causing an infectious disease happens through direct contacts with blood, body fluids, saliva, etc. [6, 7]. A virus also spreads by indirect contacts *via* natural distribution in nature, including air, water, and food, to name but a few.

7.5.2 Geographical Locations

Natural distributors, such as wind, water, air, etc., are the most important factors influencing the prediction of epidemic disease dissemination. There are two main factors, namely climatic and geodemographic, which help in determining predictive analysis based on geographical locations.

7.5.2.1 Climatic Factors

There are a number of ways through which climate can affect disease transmission. For example, local climate can heavily affect the distribution and population size of disease vectors, such as *via* sewage overflow and widespread water contamination as a result of flooding after heavy rains. The spread of pathogens from one region to another can also happen along air streams or by wind. Global warming also causes warm air, resulting in multiplication of insect vectors and increased air or water contamination [3, 4, 6].

7.5.2.2 Geodemographic Factors

Geodemographics is the study of the population classified according to geographical regions and their economic, social, and demographic characteristics. Thus, it is a classification technique to group the population residing at a specific location on the basis of their social criteria, such as sanitation and hygiene level, family structure and size, neighborhood, economic status, and education level, among others. The mismanagement of geodemographic features, which may include inappropriate waste disposal, deforestation, and widespread water contamination, affects the ecology and leads to epidemic disease dissemination [2, 4, 8].

7.5.3 Clinical Studies

Clinical studies can be performed through clinical case classification and vaccination tracking in order to determine epidemic disease outbreaks. Clinical case studies also promote machine learning techniques, which use training data for cross validation of the prediction models.

7.5.3.1 Clinical Case Classification

Epidemic disease outbreaks occur as a result of numerous factors, which must involve a large number of data gathering activities. In order to have clinical case classification, data collection techniques, such as Electronic Health Records (EHRs), are used. An EHR is a patient's electronic clinical record, which includes a patient's medical history, diagnoses records, medications, treatment plans, and laboratory and test results, etc. EHRs are used by almost all of the health organizations that are involved in tracking and storing data for the future.

Machine learning techniques, such as clustering, can be used for clinical classification. The clustering technique helps health practitioners to detect, forecast, and control the epidemic disease outbreaks by collecting the EHRs of the population [1, 2]. EHRs can be used to classify data correlations, based on disease attribute selection.

7.5.3.2 Vaccination Tracking

The clinical studies can also be conducted through vaccination tracking, which can be used to classify the vaccine type in an infected area where the vaccine was most used. In health predictive analysis, clinical studies, such as the use of EHRs, play a significant role.

7.5.4 Social Media

The social media networks have become increasingly popular in the predictive analysis of disease epidemics. The social media, such as Twitter, have become the most popular medium for information sharing.

7.5.4.1 Geo-Mapping

The metadata collected through social media networks can be used in geo-tagging, which is a process of attaching geographical identification information. Geo-tagging is helpful in tracking disease-affected areas. Social media also provides a platform from which to collect and cluster text data to identify the important disease-related terms and events discussed by users across the world.

7.6 MACHINE LEARNING ALGORITHMS FOR DISEASE EPIDEMIC PREDICTION

The choice of algorithm for the best use of machine learning depends on the type, size, and dimensionality of the data. In the literature surveyed, there are five common machine learning algorithms that have been applied for disease outbreak predictions: support vector machine, decision tree, naïve Bayes, artificial neural network, and *K*-means clustering.

7.6.1 Support Vector Machine (SVM)

SVM is the most popular supervised machine learning algorithm, which falls between two types of problems, namely classification and regression. SVM works well for both problems. However, SVM is mostly used for classification problems.

Given a set of known training data, the SVM algorithm finds an optimal hyperplane, which classifies new unknown examples. This hyperplane is a line in two-dimensional space, which divides a plane into two parts, where each class lay on either side. The data points closest to the line from both the classes are known as support vectors and the distance between support vectors and the line is called the margin. The main goal of SVM is to find the maximum margin hyperplane. The hyperplane for which the margin is maximal is called the optimal hyperplane. Figure 7.7 shows the SVM classifier that finds an optimal hyperplane.

As shown in Figure 7.7, the optimal hyperplane separates the data points with two different labels (squares and circles) on both sides of the plane in order to maximize the distances of the closest data point from both classes to the hyperplane.

7.6.2 Decision Tree

Decision tree is one of the popular supervised machine-learning algorithms for classification problems. The main goal of decision tree algorithms is to build a tree-like graph structure that predicts the value of an unknown target variable by learning simple decision rules inferred from the given set of input data features.

A decision tree is a tree where each internal node represents a feature, with each leaf node representing an outcome or a class label (categorical or continuous value) and branches representing conjunctions of features that lead to a class label. The paths from root node to leaf node represent a classification rule. Figure 7.8 shows the basic graphical representation of a decision tree for decision making.

As shown in Figure 7.8, internal nodes perform the test on data features and arrows represent the test results. Based on the test results or outcomes, leaf nodes classify the data points into two different labels or classes. The paths from the root node to the leaf node represent a decision process or classification rule. The tree generation process is recursive in nature. A tree is built by repeatedly splitting the training data set into subsets. The splitting is based on attribute selection measures, such as Information Gain, Gini Index, and Gain Ratio.

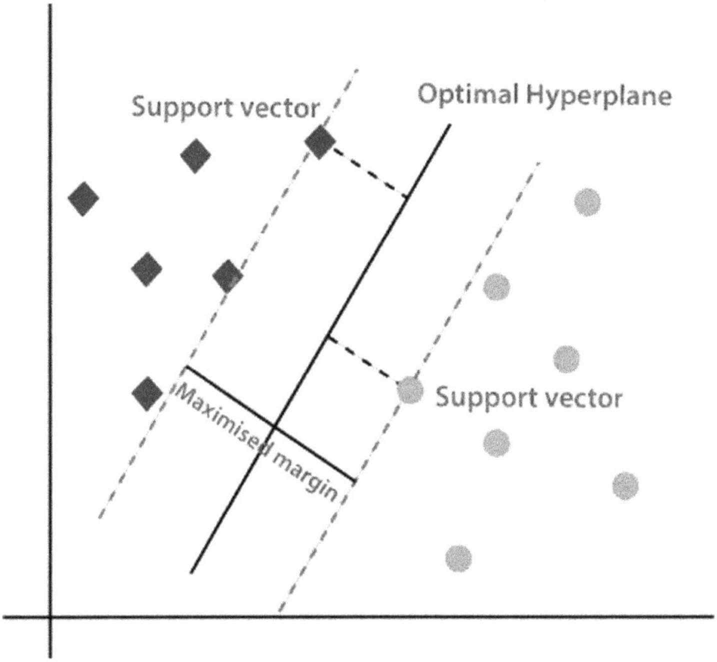

FIGURE 7.7 Optimal hyperplane, using SVM.

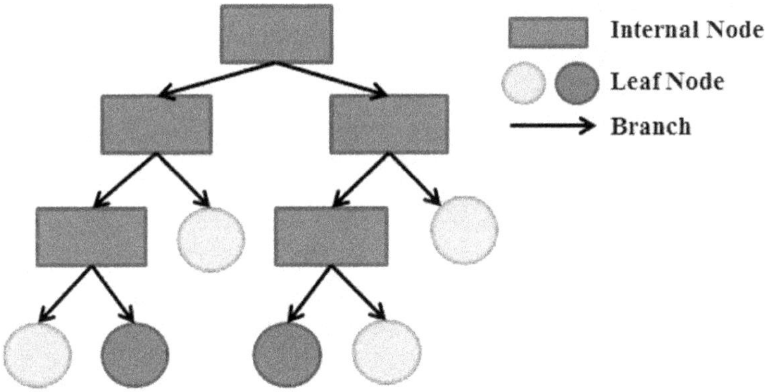

FIGURE 7.8 Decision tree.

7.6.3 Naïve Bayes

Naïve Bayes is a simple probabilistic classification supervised learning algorithm, which is based on Bayes' theorem. The theorem makes the 'naïve' assumption that the features are statistically independent, which means that the occurrence of a certain feature is independent of the occurrence of other features. The following equation represents the Bayes Theorem:

Likelihood Class prior probability

$$P(A/B) = \frac{P(B/A)P(A)}{P(B)}$$

Posterior probability Predictor prior probability

In the above equation, P(A|B) is the posterior probability or conditional probability of class A, while given a predictor B; P(A) is the prior probability of class A; P(B|A) is the likelihood which is the probability of predictor B while given class A and P(B) is the prior probability of predictor B.

Naïve Bayes calculates the posterior probability associated with each possible class and then selects the class with the highest posterior probability as the outcome of the prediction.

7.6.4 Artificial Neural Networks (ANNs)

ANNs are a type of machine learning algorithm that are modeled after the human brain and designed to recognize patterns. ANNs are computational algorithms that are intended to simulate human brain behavior, which is composed of biological units or "neurons". An artificial neural network is composed of several layers, which are made of nodes. A node is known as an artificial neuron, that is loosely modeled on a neuron in the human brain. The basic model of ANN is composed of three interconnected layers that include one input layer, one output layer, and at least one hidden layer. Each additional hidden layer increases the complexity of the training network. Figure 7.9 shows a basic ANN architecture with two hidden layers.

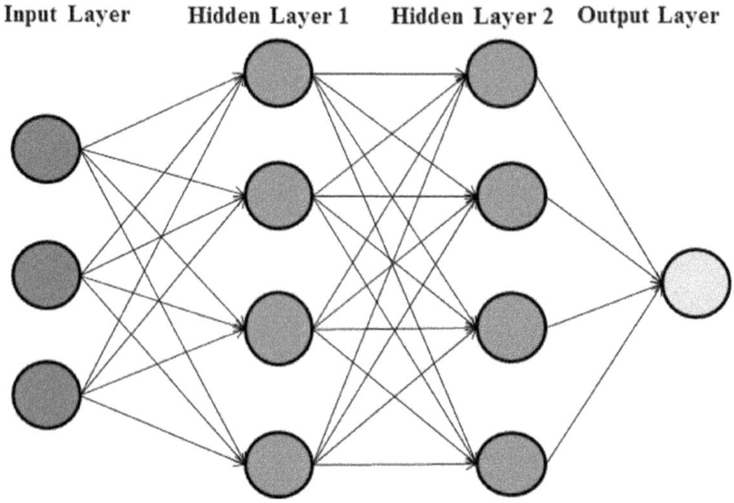

FIGURE 7.9 Graphical representation of ANN with two hidden layers.

Each neuron in the layers carries some weight. Neurons can communicate with each other in adjacent layers through activation functions. The activation function is the one that converts the weighted sum of a neuron's inputs into an output. Depending upon the type of activation function, the output can be categorical or continuous.

7.6.5 *K*-Means Clustering

K-means clustering is a simple unsupervised machine learning technique, which is considered to be one of the most-used clustering techniques. The objective of the *K*-means clustering algorithm is to identify *K* numbers of clusters in a given dataset. The data points that have similar features or attributes are grouped together to form a cluster. A cluster is defined by a center point of the cluster, which is called a centroid. Every point in a dataset is a part of the cluster whose centroid is most closely located to them. Hence, the aim of *K*-means clustering algorithm is to find *K* number of centroids. There are a number of methods for selecting *K*. Figure 7.10 shows an illustration of the *K*-means clustering algorithm.

7.7 EXISTING RESEARCH ON MACHINE LEARNING APPLICATION IN EPIDEMIC PREDICTION

Epidemic prediction and forecasting are the application of digital health epidemiology, where machine learning techniques are applied to forecast the epidemic disease outbreaks. Epidemic prediction is a process to predict the future epidemics in terms of epidemic size, peak times, and duration by using past incidence data. These kinds of predictions are crucial for efficiently preventing and controlling infectious diseases. There are a number of machine learning algorithms that have been incorporated to predict and forecast of epidemic disease outbreaks. For instance, support vector machine algorithms have been used to predict whether dengue incidence exceeded a particular threshold, using Google queries [15]. Similarly, such support vector machine algorithms have been used to predict levels of influenza-like disease 1–2 weeks before official reports, using Twitter data [16]. Gregory F. Cooper and his colleagues [17] have used a Bayesian framework, in a population with clinical

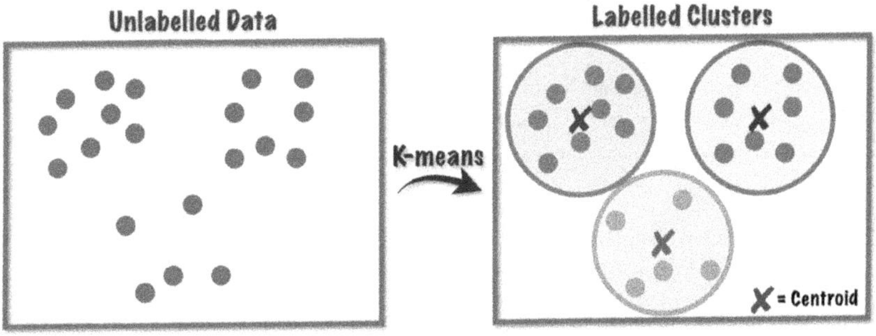

FIGURE 7.10 *K*-means clustering.

TABLE 7.1
Machine Learning Applications in Epidemiology

Objective	Reference	Machine Learning Methods
Identify risk factors for healthcare-associated ventriculitis and meningitis.	[20]	LASSO (least absolute shrinkage and selection operator) Random forest Extreme gradient booster
Influenza surveillance.	[21]	Support vector machine
Prediction of central line- associated bloodstream infection	[22]	Random forest Logistic regression
Prediction of *Clostridium difficile* recurrence.	[23]	Random forest Logistic regression
Prediction of healthcare-associated infections.	[24]	Support vector machine Gradient tree boosting
Prediction of surgical site infection	[25]	Artificial neural network Logistic regression
Prediction of daily risk of *Clostridium difficile* infection.	[26]	L2 regularized logistic regression
Prediction of central line- associated bloodstream infection and mortality.	[27]	Deep learning Logistic regression Gradient tree boosting
Prediction of surgical site infection	[28]	Naïve Bayes classifier Logistic regression
Detection and classification of healthcare-associated infections.	[29]	Naïve Bayes classifier PART algorithm (partial decision tree)
Influenza surveillance.	[30]	Least absolute shrinkage and selection operator
Real-time influenza surveillance.	[31]	Support vector machine Decision tree Regression Least absolute shrinkage and selection operator
Detection of surgical site infection	[32]	Natural language processing Bayesian network Logistic regression
Estimating costs and changes in duration of hospital stays for *Clostridium difficile* infection	[33]	Logistic regression with elastic net Regularization

diagnosis of some individuals, to showcase the epidemiological demonstration of disease outbreaks. Yilan Liao and co-workers [18] have also presented a Bayesian belief network technique to update the disease outbreak at various spatial scales in terms of virus detection rates. Table 7.1 summarizes some of the recent applications of machine learning algorithms in the field of epidemiology [19].

7.8 REAL-TIME EPIDEMIC DISEASE PREDICTION: CHALLENGES AND OPPORTUNITIES

7.8.1 Challenges

The prediction of epidemic disease outbreak gives an opportunity to predict and forecast geographic disease spread and the numbers of infected individuals to help public health practitioners and policy makers in decision making when outbreaks occur. Epidemic prediction has considerable potential in disease outbreaks but there are still some key challenges that affect their use in epidemic disease prediction [34]. These challenges are described below.

First, ensuring that updated and reliable data on basic epidemiologic parameters are available is a key challenge during a disease outbreak. The infectious disease spread depends on the current epidemiologic, demographic, and environmental situations of the community. Hence, updated and reliable data help in estimating the risk of disease spread. The frequency at which data need to be updated depends on the nature of the disease, with fast-progressing diseases requiring daily updates, whereas slow-progressing diseases may require only weekly or monthly updating of the data.

Second, uncertainty associated with predictive models is another challenge in disease outbreaks. Lack of some vital information about the emerging outbreak, the wrong choice of model, limitations of model outputs, etc., may affect the accuracy with which a model predicts the disease outbreak.

Third, addressing knowledge gaps, with regard to how end-users use prediction results, is also a challenge. There are a number of existing infectious disease predictions that are not standardized and validated. Hence, such predictions are difficult to communicate to non-scientific audiences, and predictions may fail to address results that are relevant to public health responders [35].

7.8.2 Opportunities and Advances

Despite some challenges in the prediction of disease outbreak, recent advances provide additional support and opportunities in predictive modeling of an outbreak. These are highlighted below.

First, there are a number of novel open data sources available that help in refining predictive disease models for future outbreak problems. For example, LandScan and WorldPop are programs through which demographic and geographic data are becoming more readily available.

Second, recent advances in open science approaches, such as big data, digitization, and connectivity, provide a number of valuable sources of information. Open data sharing is an opportunity that allows multiple developers to build, test, validate, and compare models.

Third, digital surveillance systems are also an emerging opportunity in epidemic prediction. These internet-based systems aim to collect and analyze the unstructured data from heterogeneous sources, such as news reports, EHRs, social media, and search engine queries. Digital surveillance systems have the capability to detect events with epidemic potential earlier than official notifications [36].

7.9 RELEVANCE OF MACHINE LEARNING TO THE NOVEL CORONAVIRUS (COVID-19) OUTBREAK

Artificial intelligence can be used for identifying, forecasting, and controlling novel coronavirus (COVID-19) outbreaks. Artificial intelligence and its applications, such as machine learning, deep learning, etc., are being used by practitioners to support the fight against the spread of COVID-19, that has affected the entire world since the beginning of 2020. Machine learning techniques, which use data-driven predictive analytics, make use of large volumes of data, can lead to deeper knowledge of COVID-19, and can help health practitioners and government policy makers to make better and faster decisions throughout the entire COVID-19 outbreak. The following are some areas where machine learning can be used to address the coronavirus outbreak issue.

7.9.1 Design and Development of Vaccines and Drugs

Machine learning can significantly speed up the development process of a vaccine without compromising quality. It can also be used to predict the structure of proteins and their interactions with chemical compounds, to facilitate the design of new drugs.

7.9.2 Predicting the Spread of Virus, Using Social Media Platforms

Machine learning prediction models can be used to interpret the public interaction data on social media platforms, such as Twitter, to assess the likelihood of novel COVID-19 infection. The model can estimate the spread of the epidemic in real time and can forecast the rate of spread in the future.

7.9.3 Diagnosing Virus Infection via Medical Images

Deep learning, a variant of machine learning, can be used in radiology to diagnose coronavirus. Radiologists are making use of deep learning-based artificial intelligence (AI) algorithms to detect COVID-19 in chest computed tomography (CT) scans and X-rays. The data from CT scans of COVID-positive patients can be used as input for machine learning models, which help doctors to diagnose the disease more rapidly and to develop a more-timely treatment plan. Similarly, models can be trained with the help of data from confirmed cases and can identify patients with the COVID-19 virus in CT scans. AI-enabled image acquisition can also help automate the scanning procedure with minimal contact with the patient in order to provide protection to the radiologist.

7.9.4 AI-Based Chatbots for Diagnosis

Chatbots are "self-triage" systems, where patients complete a questionnaire about their symptoms and medical history, and can self-identify their best course of action, given their specific symptoms, before calling a doctor or visiting a hospital. Many companies like Microsoft have created Chatbots that can determine the probability of the patient being COVID-19-positive.

7.9.5 Smartphone Application Developments

Artificial intelligence and machine learning can be used for developing smartphone applications ("apps") that can be used for advance prediction of COVID-19 disease based on symptoms, such as dry cough, fever etc. The machine learning techniques can also be used in developing mobile apps that help in contact tracing and alerting people whenever a COVID-19-positive person having such an app on his/her mobile phone comes in the vicinity of others having same app on their mobile phones. This helps in containing the further spread of disease. The AarogyaSetu mobile app, developed by the Government of India, is an example of such use of machine learning techniques in containing the spread of the COVID-19 outbreak.

REFERENCES

1. Boivin G, Hardy I, Tellier G, Maziade J. 2000. Predicting Influenza Infections during Epidemics with Use of Clinical Case Definition. *Clinical Infectious Diseases.* 31(5): pp. 1166–1169.
2. Ravì D, Wong C, Deligianni F et al. 2017. Deep Learning for Health Informatics. *IEEE J. Biomed. Health Inform.* 21(1): pp. 4–21.
3. Shope R. 1991. Global Climate Change and Infectious Diseases. *Environ. Health Perspec.* 96: pp. 171–174.
4. Shuman EK. 2010. Global Climate Change and Infectious Diseases. *N. Engl. J. Med.* 362(12): pp. 1061–1063.
5. Sadilek A, Kautz H, and Silenzio V. 2012. Predicting Disease Transmission from Geo Tagged Micro-Blog Data, in *Proceedings of the Twenty-Sixth AAAI Conference on Artificial Intelligence.* AAAI Press: Toronto, ON, Canada. p. 136–142.
6. Andrick B, Clark B, Nygaard K et al. 1997. Infectious Disease and Climate Change: Detecting Contributing Factors and Predicting Future Outbreaks, in *Geoscience and Remote Sensing. IGARSS '97. Remote Sensing - A Scientific Vision for Sustainable Development.* IEEE International.
7. Masuda N. and P. Holme. 2013. Predicting and Controlling Infectious Disease Epidemics Using Temporal Networks. F1000 Prime Reports. 5: p. 6.
8. Kimura Y, Saito R, Tsujimoto Y et al. 2011. Geodemographics Profiling of Influenza A and B Virus Infections in Community Neighborhoods in Japan. *BMC Infect. Dis.* 11(1): p. 36.
9. Big data Analytics and Predictive Analytics - Predictive Analytics Today (2018). [Online]. Available: http://www.predictiveanalyticstoday.com/big-data- analytics-and-predictive-analytics/.

10. The App Solutions, How Predictive Analysis is changing Healthcare Industry (2019). Available: https://medium.com/@TheAPPSolutions/how-predictive-analytics-is-changing-healthcare-industry/.
11. Tom M. 1997. *Machine Learning*. New York: McGraw Hill.
12. Samuel AL. 1959. Some Studies in Machine Learning Using the Game of Checkers. *IBM J. Res. Dev.* 3(3): pp. 210–229.
13. Duda RO, Hart PE, Stork DG. 2012. *Pattern Classification*. 2nd Ed. Hoboken, NJ: John Wiley & Sons. 517 p.
14. George DB, Taylor W, Shaman J et al. 2019. Technology to Advance Infectious Disease Forecasting for Outbreak Management. *Nat. Commun.* 10: p. 3932.
15. Althouse BM, Ng YY, Cummings DAT. 2011. Prediction of Dengue Incidence Using Search Query Surveillance. *PLoS Negl. Trop. Dis.* 5(8): p. e1258.
16. Signorini A, Segre AM, Polgreen PM. 2011. The use of Twitter to Track Levels of Disease Activity and Public Concern in the U.S. During the Influenza A H1N1 Pandemic. *PLoS One.* 6(5): e19467.
17. Cooper G, Villamarin R, (Rich) Tsui F, Millett N, Espino J and Wagner M. 2015. A Method for Detecting and Characterizing Outbreaks of Infectious Disease from Clinical Reports. *Journal of Biomedical Informatics.* 53: pp. 15–26,
18. Liao Y, Xu B, Wang J and Liu X. 2017. A New Method for Assessing the Risk of Infectious Disease Outbreak. *Sci. Rep.*, 7: 1–12.
19. Roth JA, Manuel Battegay, Fabrice Juchler, Julia E. Vogt, Andreas F. Widmer. 2018. Introduction to Machine Learning in Digital Healthcare Epidemiology. *Infect. Control Hosp. Epidemiol.* 39(12): 1–6.
20. Savin I, Ershova K, Kurdyumova N, et al. 2018. Healthcare-Associated Ventriculitis and Meningitis in a Neuro-ICU: Incidence and Risk Factors Selected by Machine Learning Approach. *J. Crit. Care.* 45: pp. 95–104.
21. Allen C, Tsou M-H, Aslam A, Nagel A, Gawron J-M. 2016. Applying GIS and Machine Learning Methods to Twitter Data for Multiscale Surveillance of Influenza. *PLoS One.* 11: p. e0157734.
22. Beeler C, Dbeibo L, Kelley K, et al. 2018. Assessing Patient Risk of Central Line-Associated Bacteremia via Machine Learning. *Am. J. Infect. Control.* 46: pp. 986–991.
23. Escobar GJ, Baker JM, Kipnis P, et al. 2017. Prediction of Recurrent *Clostridium difficile* Infection using Comprehensive Electronic Medical Records in an Integrated Healthcare Delivery System. *Infect Control Hosp Epidemiol.* 38: pp. 1196–1203.
24. Ehrentraut C, Ekholm M, Tanushi H, Tiedemann J, Dalianis H. 2018. Detecting Hospital-Acquired Infections: A Document Classification Approach Using Support Vector Machines and Gradient Tree Boosting. *Health Informatics J.* 24: p. 24.
25. Kuo P-J, Wu S-C, Chien P-C, et al. 2018. Artificial Neural Network Approach to Predict Surgical Site Infection After Free-Flap Reconstruction in Patients Receiving Surgery for Head and Neck Cancer. *Oncotarget.* 9: pp. 13768–13782.
26. Oh J, Makar M, Fusco C, et al. 2018. A Generalizable, Data-Driven Approach to Predict Daily Risk of *Clostridium difficile* Infection at Two Large Academic Health Centers. *Infect. Control Hosp. Epidemiol.* 39: pp. 425–433.
27. Parreco JP, Hidalgo AE, Badilla AD, Ilyas O, Rattan R. 2018. Predicting Central Line-Associated Bloodstream Infections and Mortality Using Supervised Machine Learning. *J. Crit, Care.* 45: pp. 156–162.
28. Sanger PC, van Ramshorst GH, Mercan E, et al. 2016. A Prognostic Model of Surgical Site Infection Using Daily Clinical Wound Assessment. *J. Am. Coll. Surg.* 223: pp. 259–270.

29. Gómez-Vallejo HJ, Uriel-Latorre B, Sande-Meijide M, et al. 2016. A Case-Based Reasoning System for Aiding Detection and Classification of Nosocomial Infections. *Decision Support Syst.* 84: pp. 104–116.
30. Lu FS, Hou S, Baltrusaitis K, et al. 2018. Accurate Influenza Monitoring and Forecasting Using Novel Internet Data Streams: A Case Study in the Boston Metropolis. *JMIR Public Health Surveill.* 4: p. e4.
31. Santillana M, Nguyen AT, Dredze M, Paul MJ, Nsoesie EO, Brownstein JS. 2015. Combining Search, Social Media, and Traditional Data Sources to Improve Influenza Surveillance. *PLoS Comput. Biol.* 11: p. e1004513.
32. Sohn S, Larson DW, Habermann EB, Naessens JM, Alabbad JY, Liu H. 2017. Detection of Clinically Important Colorectal Surgical Site Infection Using Bayesian Network. *J. Surg. Res.* 209: pp. 168–173.
33. Pak TR, Chacko KI, O'Donnell T, et al. 2017. Estimating Local Costs Associated with *Clostridium difficile* Infection Using Machine Learning and Electronic Medical Records. *Infect. Control Hosp. Epidemiol.* 38: pp. 1478–1486.
34. Desai AN, Kraemer MU, Bhatia S, et al. 2019. Real-time Epidemic Forecasting: Challenges and Opportunities. *Health Security* 17(4): 268–275.
35. Doms C, Kramer SC, Shaman J. 2018. Assessing the Use of Influenza Forecasts and Epidemiological Modeling in Public Health Decision Making in the United States. *Sci Rep.* 8(1): p. 12406.
36. Milinovich GJ, Williams GM, Clements AC, Hu W. 2014. Internet-Based Surveillance Systems for Monitoring Emerging Infectious Diseases. *Lancet Infect. Dis.* 14: pp. 160–168. doi: 10.1016/S1473-3099(13)70244-5.

8 Machine Learning–Based Case Studies for Healthcare Analytics

Electronic Health Records, Smart Health Monitoring, Disease Prediction, Precision Medicine, and Clinical Support Systems

T. Kokilavani and T. Lucia Agnes Beena

CONTENTS

8.1 INTRODUCTION

Digital data is considered to be the new oil for the development of data science. According to Forbes, by the end of 2025, digital data will increase significantly up to 4.4 ZB (4.4×10^{21} bytes). Each human being in the world is generating, on average, 1.7 MB digital data every second. Machine learning plays an important role in enhancing the performance of data analytics. Machine learning is based on Artificial Intelligence, which gives the system the capability to itself learn from collected data, without the need for explicit programming. The main concern of machine learning is the accuracy and performance of the system. The machine learning algorithms learn patterns from data over time. These algorithms are trained with training data to produce accurate models to extract useful information from the training data. When a model is developed, it will provide an output based on the prior training it was given with the training dataset. Machine learning algorithms fall under one of four categories, namely supervised learning, unsupervised learning, semi-supervised learning, or reinforcement learning. The success of the machine learning system depends on the type of algorithm used. The machine learning algorithms control the searching process to build the knowledge structures.

Machine learning techniques are used for numerous applications around the globe. The healthcare industry employs these algorithms for predicting the presence/absence of various conditions, like heart problems, brain tumors, increased/decreased blood pressure, high blood sugar, obesity, etc. Decision Trees are a form of machine learning algorithm, which are used for identifying the presence of brain tumors. According to the World Health Statistics 2012, one in three of the deaths each day occur because of heart problems. Predictive models for heart problems developed by various researchers extract the hidden information from heart problem datasets. Heart disease prediction systems help medical professionals and doctors to predict the problem accurately and efficiently.

The Naïve Bayes Classifier (NBC) model has been trained to mine the electronic health records of the patients in Boston Hospital to identify the suicidal behavior of patients. Electronic health records are used for early detection of the prevalence of this behavior. Machine learning techniques are widely used in clinical research, employing electronic health records. The Support Vector Machine (SVM) model is used for classifying patients with similar diseases. A new model has been developed, using SVM, for efficient monitoring of patients admitted to intensive care units (ICUs). This model outperforms the existing model by avoiding the generation of false positives. Because of differences in lifestyle habits and environmental factors, humans develop different diseases. Prediction of these diseases at an early stage becomes a great challenge for physicians. The Convolutional Neural Network (CNN) is a deep learning algorithm, which can be used for the exact prediction of disease from the disease symptom dataset at an early stage. Structured and unstructured data collected from hospitals are used for disease prediction, using machine learning methods like Decision Tree and Random Forest algorithms. They are also used for segregating the data, based on disease patterns. The principal theory behind precision medicine is to design a model for diagnosing, identifying disease risk, and predicting and monitoring response to treatment, based on the multidimensional biological datasets collected from various patients. Clinical decision support systems follow a two-layer (disease–symptom) model which may use Naïve Bayes algorithms to calculate the probability of patients suffering from the disease. This chapter covers different case studies, such as the health monitoring system, disease prediction, precision medicine, and clinical support systems for better prediction of diseases, using machine learning methods to provide proper healthcare treatments for people.

8.2 ELECTRONIC HEALTH RECORDS

The evolution of technology, like cloud computing, big data, Web 2.0, Internet of Things, and 5G networks, are supporting patient trust in healthcare systems. Digitization of healthcare services transformed the paper-based medical/health data into the paperless format of electronic health/medical records. Electronic Medical Records (EMRs) and Electronic Health Records (EHRs) are practicalities in structuring the smart health systems [1]. EHRs in healthcare systems improve efficiency in healthcare service delivery and increase patient safety, amplifying access to healthcare services, and reducing the costs of medical treatment [2]. EMRs are medical data in a digital format, that is easy to store, update, and exchange between healthcare institutions elsewhere. The EHRs hold a patient's demographic details, treatment notes, critical signs, health check histories, diagnoses, pills, vaccination dates, allergies, radiology images, test results, clerical data, and invoice data. The EHR approach is created to communicate information between healthcare providers, like specialists, pharmacies, laboratories, medical imaging facilities, and treatment centers.

EHR is the result of combining healthcare and data science. Machine learning and EHRs form the best blend for precise reporting of results and for efficient actions when treating the patient [3]. The EHR data are updated in real time and shared with all those involved in caring for the patient. The machine learning algorithms help to provide improved predictive analysis. Hence, the availability of EHRs encourages the doctors to achieve more accurate and rapid diagnoses. The EHRs secure the patient details using encryption and have data protection routines. Only authorized users can access the EHRs. In the United States, the Health Information Technology for Economic and Clinical Health Act (HITECH), the Health Insurance Portability and Accountability Act (HIPAA), and the Personal Information Protection and Electronic Documents Act (PIPEDA) [4] enable EHR technology to provide data protection and information security for the patients' data. The availability of data in EHR with essential security encourages the data scientists to apply machine learning algorithms for medical predictions.

Researchers used different machine learning methods, such as supervised (Logistic Regression, Support Vector Machines, Random Forest, Artificial Neural Networks, and XGBoost) [5, 6], semi-supervised (Label Learning method) [7], and unsupervised (Latent Dirichlet Allocation) [8] models for predicting healthcare issues, with the support of EHRs. Logistic Regression, Random Forest, Support Vector Machine, Artificial Neural Network, Gradient Boosting, Linear Nearest

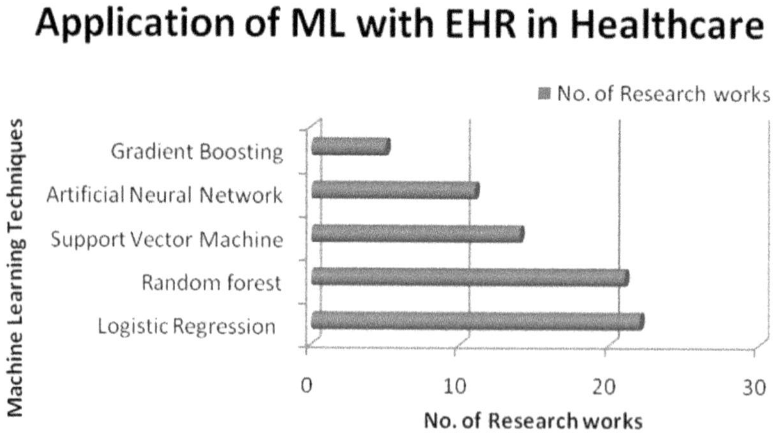

FIGURE 8.1 Adoption of machine learning in healthcare research.

Neighbor, *K*-Nearest Neighbor, Stochastic Gradient Descent, and the Cox model are mostly used in state-of-the-art systems. Figure 8.1 shows the frequency of adoption of various machine learning techniques in research papers [5, 9].

8.2.1 Supervised Machine Learning with EHR in Healthcare

Hospitals in wealthy nations maintain their patient data (as EHRs) in unstructured data formats. Electronic data availability enables the application of machine learning to diagnosis, the prediction of the severity of the disease, and recommendation of appropriate treatments. Among the different machine learning algorithms, supervised machine learning is used to identify whether the patients are diseased or healthy, evaluating the costs for a period of hospital stay, and predicting various infections [5, 10].

In order to improve the concern for the patients, Healthcare Social Networks (HSNs) are used. HSNs sometimes present erroneous information and inappropriate interpretations which affect the patient's health and mind. Though the medical practitioners like to exchange information through HSNs, they will not be able to read and validate the contents of the patient's records. In this case, supervised machine learning can identify the critical post submitted by patients and alert the practitioners to validate the post. Lorenzo et al. [11] employed Linear, Bayesian and Support Vector Machine classifiers to recognize dangerous posts and to notify the doctors to be involved in the discussion.

8.2.2 Semi-Supervised Machine Learning with EHRs in Health Care

EHR details can be applied to predict the probability of future occurrence of diseases like diabetes, autism, rheumatoid arthritis, and inflammatory bowel. In addition, EHRs are a potential instrument for recognizing the relations between multiple phenotypes and genomic markers. The main objective of designing EHRs was improved patient care, record keeping, and invoicing. So, EHRs were not initially used to extract phenotypic particulars of a patient, involving physical investigation of medical charts. Recently, efforts have been taken to integrate EHRs and a genomics network (eMERGE) [8] for determining status by applying phenotyping algorithms. These phenotyping algorithms are in the developmental stage, using small training datasets that follow gold standard labels obtained by means of chart assessments. Many methods were attempted to achieve EHR phenotyping. When applying supervised machine learning approaches, extensive training sets are needed to provide the generalized algorithms, when the independent variable is high dimensional in nature. Unsupervised learning approaches can be alternative approaches. But even standard unsupervised techniques are unsuccessful when the dimension of the independent variable is high. So, it is better to adopt semi-supervised techniques for EHR phenotyping.

A study was carried out [12] using Partner's Healthcare Systems to classify patients with rheumatoid arthritis (RA), with the help of EHR data. The EHR characteristics include the number of ICD-9 (The International Classification of Diseases Clinical Modification, 9th Revision) codes and medical terms related to RA extracted through Natural Language Processing (NLP). Among the 46,114 patients in the study who possessed at least one ICD-9 code, a random sample of 435 patients had rheumatoid arthritis, identified by rheumatologists through physical chart review. In this study, prior knowledge was included in the high-dimensional sparse estimator, reducing the required number of labeled samples. This Prior Adaptive Semi-supervised (PASS) method is also powerful in detecting defective prior information. When compared to existing methods, it was found that the performance of PASS was superior.

8.2.3 Unsupervised Machine Learning with EHR in Health Care

The EHR data are unstructured and contain medical information like ICD-10 classes, results of laboratory assessments, and disease indicators exhibited by patients. The data contain missing values which cause problems during analysis, using computing systems. If these difficulties are overcome, it will help the medical practitioners to a great extent. Unsupervised machine learning techniques are utilized by researchers to extract useful insights from the EHR data.

Clustering, one of the important unsupervised machine learning methods, is adopted to identify patient subgroups. The major objective of clustering is to identify groups from unlabeled data having similar characteristics. It can be used in data labeling before the application of supervised learning. In data visualization and analysis, it is used in the pre-processing phase. Wang et al. [8] explored the EHR data, enhancing Latent Dirichlet Allocation (LDA) techniques to the Poisson Dirichlet Model (PDM), that determines the latent disease groups. Elin [13] used *K*-means and Agglomerative Hierarchical algorithms to conduct experiments on raw data and on imputation data. The silhouette value (a measure of how similar an object is to its own cluster (cohesion), compared with other clusters [separation]), obtained using cosine and Euclidean distance measures was the evaluation index. Fair distance interpretation is given by the cosine measure. The data imputation improved the clustering results. Superior quality clusters were formed by Hierarchical Clustering than the *K*-means Clustering.

8.3 SMART HEALTH MONITORING

In recent years, the healthcare sector has faced major challenges. Huge numbers of people are prone to diseases such as asthma, arthritis, Alzheimer's disease, cancer, diabetes, dementia, or cardiovascular diseases, which require regular diagnosis, observation, and physician advice. The goal of the health monitoring system proposed by Rejab [14] is to design a common model for clustering patients suffering from related diseases. The Incremental Learning with Support Vector Machine (ISVM) and OnLine Algorithm Support Vector Machine (LASVM) techniques, with *K*-prototypes of Support Vector Machines, were used for clustering patients having similar diseases. The real clinical databases were analyzed and found that the K Prototype OnLine Algorithm Support Vector Machine (KP-LASVM) technique gave the best performance. Isaac et al. discussed a smart healthcare system [15] for assisting weight management and obesity, using Internet of Things (IoT) devices and machine learning algorithms. This system collected the data in real time through smart devices and wearable devices. The patient's important biomedical attributes are considered for analysis. The machine learning algorithms detect the significant attributes and support the patients, regarding weight loss. Samik et al. proposed a remote smart health examining system [16], using nonintrusive sensors to collect data on the patient's health parameters, like blood pressure, ECG, and respiration. It gives the autonomy for the user to send the data to the Cloud. This system is appropriate for elders who are in homes or in rural areas, for self-health monitoring. The results of the observations can be forwarded by the elders themselves to doctors for analysis, using the IoT ecosystem.

The general framework for the smart health monitoring system is shown in Figure 8.2. Values for the vital medical parameters appropriate for the disease in question can be collected through wearable devices. Based on the medical parameters, the comparison is made with the smart medical scale and the result is transferred to the smart phone, which is connected and synchronized with the wearable devices. The data collected by the smart phone is sent to the device provider and stored in the Cloud. The smart health monitoring platform can receive and analyze each patient's data from the Cloud to detect possible criticality and recommend medical services for the patient.

An open-source IoT application, ThingSpeak, and Application Programming Interfaces (API) are used to store and retrieve data from sensors, using the Hyper Text Transfer Protocol (HTTP) protocol over the internet. The data are transmitted to ThingSpeak using the Representational State Transfer (REST) API. ThinkSpeak uses HTTP requests, along with the API key secure communication, and to embed the channel's content that are to be analyzed on a personal website [16]. Therefore, the user can visualize the data in the web browser of a smartphone or a computer wherever they are. Furthermore, the doctors can connect to the channel and

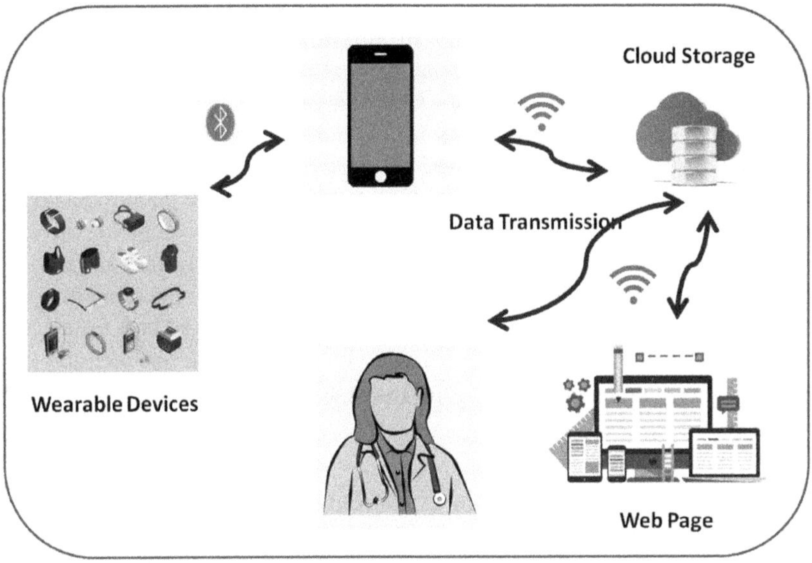

FIGURE 8.2 General framework for smart health monitoring system.

recommend appropriate medication. An automated system can use Weka API and the J48 machine learning algorithms to identify the critical attributes, while RuleML and Apache Mahout can be applied for medical recommendations [15].

8.4 DISEASE PREDICTION

Due to environmental factors and the ever-growing population, many human beings are each affected by a number of diseases nowadays. Because of inappropriate lifestyle habits, unbalanced sleep routines, unhealthy diet, and inadequate physical exercise, people face lots of problems, which lead to chronic disease. The accurate prediction of such diseases is particularly important for treatment, which is a challenging task for doctors. Also, it is important to predict the disease at an early stage, to optimize the response to treatment. Medical data are increasing every day, leading to the generation of big data. These demands of the healthcare sectors have made them adopt computer-based services. Machine learning algorithms are considered as a subfield of Artificial Intelligence [17]. According to Arthur Samuel [18] the ability to design and develop algorithms without the help of human beings is termed as machine learning. Automatic learning through expressive models is provided by machine learning. These algorithms can be used for fast and accurate prediction of disease. Some algorithms work only on structured data, whereas Convolutional Neural Network (CNN) machine learning algorithms uses both unstructured and structured datasets from hospitals for the classification of diseases. Data mining and machine learning algorithms go hand-in-hand for prediction, modeling, and decision making.

Based on the health data and treatment history of patients, machine learning strategies have been increasingly applied for the prediction of diseases in recent decades [19]. Huge volumes of data generated by the healthcare sector are difficult to store and process manually by humans. Machine learning algorithms play a major role in handling these health datasets with high-quality analysis and at minimal cost. These algorithms fall under one of three categories, namely tree-based, probability-based, or rule-based algorithms. After collecting data from different patients, data pre-processing is carried out to remove irrelevant data, thereby reducing the size of the dataset. The system cannot take a correct decision if the volume of data is too high. Thus, different types of algorithms are developed to extract useful relevant information from the historical data. Once the prediction model is developed for a particular disease, it is tested using the test data to determine the accuracy of the model. The developed model can be optimized again by considering disease-based parameters [20].

8.4.1 Predicting the Presence of Heart Diseases

A statistics released by World Health Organization (WHO) during 2020 says that number one cause for death is cardiovascular diseases. According to the estimation of WHO, 17.9 million people die due to heart diseases each year. Various research is in progress to predict heart disease, by extracting the hidden information from individual patient's datasets. The prediction system for heart disease helps doctors to predict the presence of the illness more accurately and more efficiently, to make the best medical decisions. Multiple and multivariate regression models are suitable for heart disease prediction. Multiple regression models refer to one dependent variable and multiple independent variables, whereas multivariate regression models refer to multiple dependent and multiple independent variables [18].

Bhuvaneeswari et al. [21] has stated that heart disease is the main reason for the increasing death rate among human populations. Identifying the existence of heart disease is considered to be an important focus in the healthcare industry. The authors proposed a machine learning procedure J48 Classifier as a prediction model. They used Correlation-based Feature Selection (CFS) for the selection of features by which to improve the prediction results.

According to a research study carried out by the Registrar General of India and the Indian Council of Medical Research, 25% of the deaths in people within the age category 25 to 69, result from heart problems [22]. Figure 8.3 shows the prediction result of this research study.

The existence of a particular disease in a human being is identified by various parameters. In the case of cardiovascular disease (or "heart disease"), some of the parameters used to identify the presence of the illness are high blood pressure, blocked blood vessels, high blood glucose level, and high blood cholesterol (specifically, high low-density lipoprotein cholesterol). Machine learning algorithms are found to be effective in identifying the symptoms and predicting heart problems from the large volume of data generated from the healthcare sector. Shubhankar Rawat [23] has designed a predictive model in which the Cleveland Heart Disease

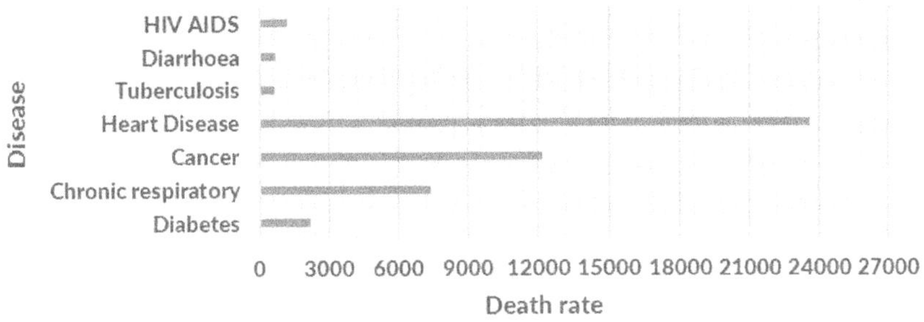

FIGURE 8.3 Predicted death rate by 2030.

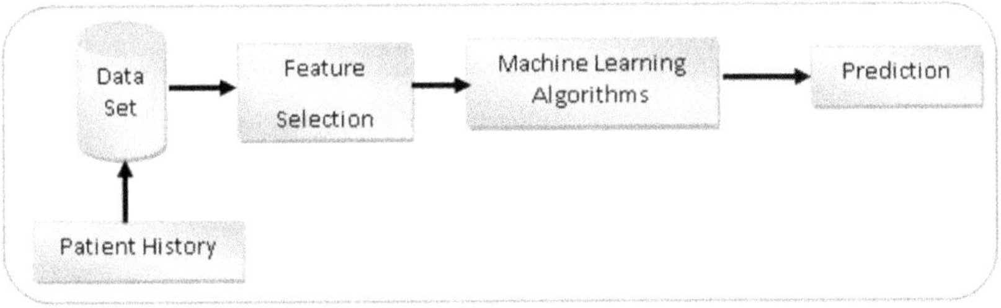

FIGURE 8.4 Predictive model for heart disease.

dataset is used to collect data for the relevant parameters. Most researchers use this dataset for designing a predictive model for heart disease. The dataset contains 76 parameters, out of which the author used 14 major factors to identify the presence of heart disease. The factors he selected were age, sex, angina, resting blood pressure, serum cholesterol, fasting blood sugar, resting ECG, maximum heart rate achieved, exercise-induced angina, peak exercise ST segment (The ST segment is the flat, isoelectric section of the ECG between the end of the S wave (the J point) and the beginning of the T wave), ST depression induced by exercise relative to rest, number of major vessels colored by fluoroscopy, thalassemia, and diagnosis of heart disease. Karan Bhanot [24] used some machine learning techniques like Support Vector Classifier, *K*-Neighbors Classifier, Random Forest Classifier, and Decision Tree Classifier. This author compared the score of these four techniques from datasets of patients affected by heart disease. The scores were taken after proper training and testing. Among the four classifier models, *K*-Neighbors Classifier gave the highest score. The design of a prediction model for heart disease, using any of these machine learning techniques, may be achieved as shown in Figure 8.4.

Different researchers have proposed different machine learning strategies to identify the presence of cardiovascular disease in people based on their medical history. Table 8.1 shows the accuracy achieved from applying these techniques. Some of the researchers reported the accuracy achieved before and after feature selection. The research study shows that better results were achieved after proper feature selection. Figure 8.5 shows the graphical representation of the results.

TABLE 8.1
Accuracy of Various Machine Learning Techniques

S. No.	Methodology	Accuracy
1	Logistic Regression Classifier [25]	77%
2	Multilayer Perceptron and Support Vector Machine [26]	80.41%
3	Integrated Fuzzy Neural Network and Artificial Neural Network [27]	87.4%
4	Naïve Bayes [28]	86.12%
5	Decision Tree Classifier [28]	80.4%
6	Three-phase Artificial Neural Network model [29]	88.89%
7	Artificial Neural Network ensemble-based predictive model [30]	89.01%
8	Least Absolute Shrinkage and Selection Operator (LASSO) model with feature selection [31]	89%
9	Majority vote with NB, BN, RF and MP [32]	85.48%

Note: Naïve Bayes (NB), Bayes Net (BN), Random Forest (RF), Multilayer Perceptron (MP).

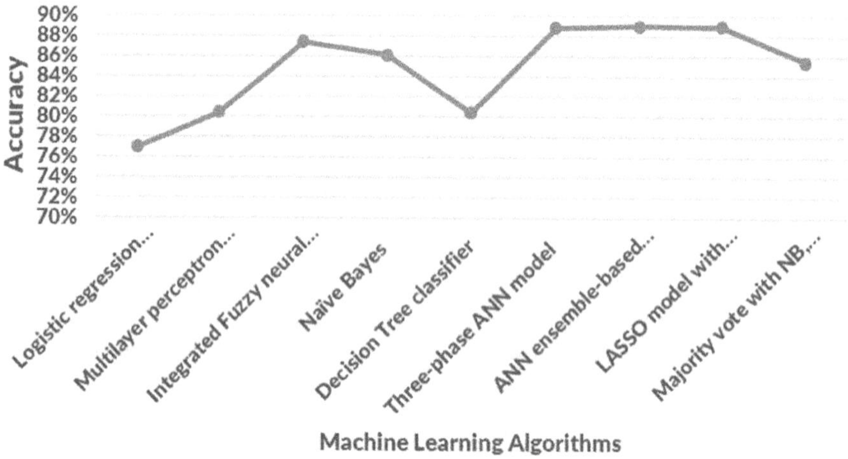

FIGURE 8.5 Accuracy of machine learning techniques.

8.5 PRECISION MEDICINE

Precision medicine is not a new terminology to the healthcare industry. It has been used in the healthcare domain for many years. When a patient is in need of blood transfusion, only donor blood of the ABO blood group which matches with that of the patient's blood is appropriate. This personalized approach, which focuses on the individual patient's history, has been extended to other areas of medicine in the subsequent years [33].

Precision medicine, also called personalized medicine, is a novel model which has the potential for highly appropriate diagnosis, treatment, and care for specific individuals. The medical research community has focused significant attention on precision medicine in recent years. The focus on precision medicine is necessary because of three fundamental factors, namely physical composition of gene types (inherited genes), population size, and statistical investigation [34]. Because of dissimilarities in biological characteristics, a subset of people may differ in their disease risk, prediction, and response to treatment. The fundamental theory of precision medicine is to identify those people who are particularly vulnerable to the disease, to achieve early detection of the disease. Machine learning algorithms can be used to train the available multidimensional datasets to capture the variation in the patients' biological factors. Current technologies, like High-Performance Computing (HPC), are needed for the execution of a precision medicine path, which would transform the healthcare industry [35].

Hereditary details and electronic health records (EHRs) of patient data has improved the doctor's approach to individual patients, allowing treatments to be customized to their specific needs. Based on the medical history of inherited traits, genetic characters and data collected from wearable devices, treatments, or medicines can be intended for small groups, instead of huge populations, a strategy which is termed precision medicine. Dr. Elizabeth Krakow has used machine learning-based precision cancer treatments. Dr. Krakow studied the data from blood cancer patients and proposed that the appropriate order of treatments, like stem cell transplantation or several sequences of chemotherapy and immunotherapy, may vary because, each time, the appropriate choice of treatment depends on each patient's current health condition. So, she has designed a machine learning algorithm to predict the exact treatment order at any specific time. The proposed algorithm simulates the treatment and predicts the results by matching their profiles with the patient's past data [36]. Figure 8.6 shows how precision medicine gives better results than the standard approach to diagnosis and treatment.

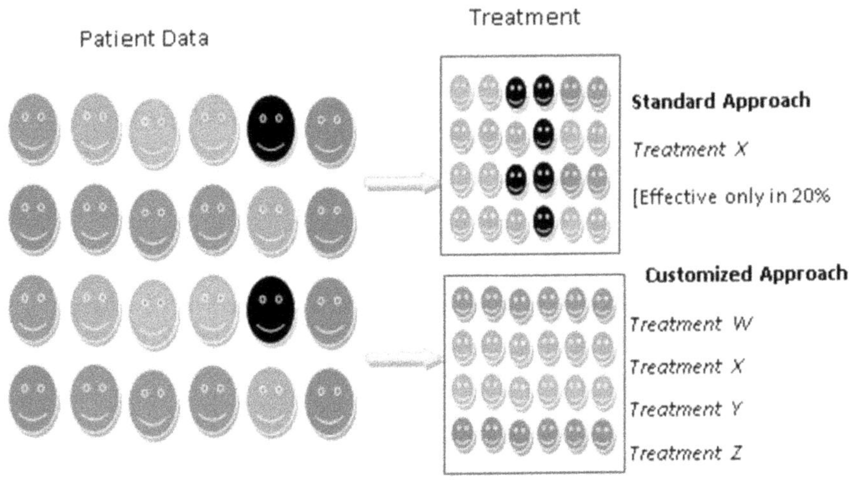

FIGURE 8.6 Precision medicine approach for treatment.

In the early stage of childhood, neurodevelopmental disorders will begin which affect various characters, like language, cognitive, social, behavioral and psychomotor control. Different disorders, such as autism, intellectual disability, movement disorders, tic disorders, and attention deficit hyperactivity disorder, fall under the neurodevelopmental disorder category. Some developmental disorders, such as schizophrenia, arise during the intermediate stage of development. Machine learning algorithms, like supervised learning, unsupervised learning, and semi-supervised learning, are capable of analyzing the biological datasets and providing predictive models for identification and treatment of all the disorders happening in either the childhood or intermediate stages. Using supervised learning algorithm, the prediction model is designed with the help of training data. This technique matches input to an output, using known input–output pairs given during training. The unsupervised learning technique maps the input with the output by discovering patterns from the datasets. It does not have labeled training data as supervised learning does. In a semi-supervised learning algorithm, a mixture of known and unknown training data is used to identify the input–output matches [34].

Precision medicine also involves identifying strategies for multifactorial illnesses and finding alternative solutions for treatment. Most of the precision medicine research involves unsupervised learning, which requires finding patterns in the medical data, whereas supervised learning involves predictions [37]. Microsoft Azure uses machine learning algorithms to aid in most of the innovative research and development in precision medicine.

8.6 CLINICAL DECISION SUPPORT SYSTEM

A Clinical Decision Support System (CDSS) is used to improve the healthcare industry by enhancing clinical decisions, using the available medical knowledge and the patient's history. CDSS consists of software that matches the computerized medical knowledge with the individual patient's details, which directly helps in the decision-making process. CDSS is implemented through computers, smartphones, tablets, and biometric wearable devices to collect patient data, which are stored as electronic health records. Two categories of CDSS are used in the healthcare field, namely knowledge-based and non-knowledge-based classes. Rules are formed in knowledge-based systems, which retrieve data to execute the rule and, based on the rule, an action or output is generated. Based on patients, practice, or literature, rules are formed. Instead of following the expert's medical knowledge, non-knowledge-based systems use machine learning techniques to take decisions [38]. Some of the important characteristics of CDSS are: (i) Organized steps, to produce new data, which helps with decision making, (ii) Feedback loops and chronological repetitions, (iii) Merging data, (iv) Using cognitive skills and knowledge information processed to take decisions, and (v) Data reuse [39].

The data collected from wearable devices are complicated and hence CDSS is needed, which can integrate patient data from heterogeneous sources, such as the hereditary information of patients, laboratory test results, and current health information. CDSS helps physicians to recommend identification of a list of diseases or to diagnose a particular disease which may lead to other diseases. In this way, machine learning algorithms are used to determine the correlation existing among various diseases [40]. The application of machine learning algorithms to a huge medical dataset may lead to innovative trends, which may be useful for doctors. The application of machine learning algorithms to the healthcare industry has improved the quality and safety considerably. In particular, the safety of high-risk field neurosurgery has been improved [41].

CDSS consists of two components, namely domain knowledge (medical knowledge) and decision making (diagnosis). It is used in the healthcare domain to improve treatments for individuals and to reduce the timing taken for decision making, thereby reducing the costs and clinical error rates. CDSS can be used to diagnose many diseases [42].

8.6.1 Smart CDSS Architecture

Figure 8.7 shows the architecture of smart CDSS. Smart CDSS is used to diagnose a disease and it also helps the doctors to carry out an effective follow-up [43].

- The diagnosis and health issue detection can be interlinked with a mobile device, wearable device, or smart watch, making use of edge computing principles, or at a local computing environment or home/hospital server.

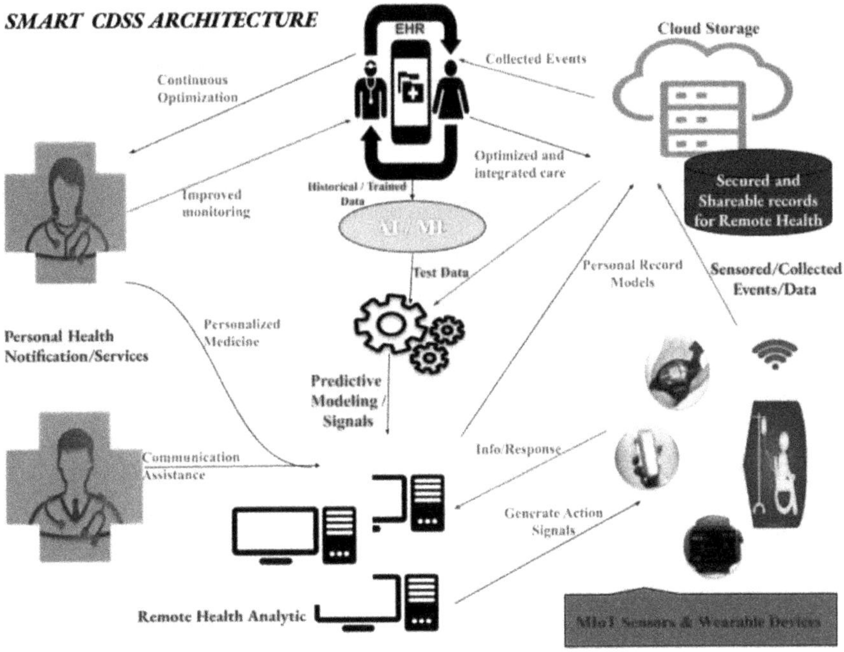

FIGURE 8.7 Clinical Decision Support System architecture.

- The MIoT (Medical Internet of Things) analysis and mining can better detect potential health problems and allow storage remotely or on the Cloud.
- From the data collected from individual mobile and wearable devices, Cloud computing components/machine learning predictive models provide efficiency and minimal latency in the discovery of dangerous medical conditions that require rapid action.
- The CDSS provides health systems suitable for an individual's welfare, where the persons concerned can be provided with treatment customized to their requirements.
- The system should be able to collect and analyze huge volumes of continuous and heterogeneous health data from smart devices used for MIoT.
- The data analysis results and predictive signals are given as input to patient smart devices, doctors, healthcare specialists, and medical organizations, through customized visual analytics and dashboard applications.

Jiang et al. [44] proposed a three-layer CDSS model, which identifies symptoms based on properties and which diagnoses a disease based on symptoms. They use a Naïve Bayes algorithm which repetitively computes the probability of the presence of a disease in patients based on multiple symptoms. This model applies the Naïve Bayes algorithm to decrease the dependencies between features. This model was assessed by two experienced physicians. By comparing the results with the previously proposed two-layer model, it was shown that this algorithm improves the accuracy of prediction and also recommends suitable treatments to the doctors.

8.7 KEY CHALLENGES

Machine learning research in healthcare is growing rapidly, with potential applications being established across various disciplines of medicine. Still, there are various limitations to applying machine learning to healthcare. The major restrictions are as follows:

- It is challenging to find appropriate datasets to analyze the problem and to find scientists knowledgeable both in machine learning techniques and in the healthcare sector.
- Medical data volumes are huge and collected from medical imaging archival systems, EHRs, pathology systems, insurance databases, and electronic prescribing tools, which are very difficult to bring together. Integrating the data to a common format creates another challenge.
- In healthcare, answering casual questions causes a number of challenges in designing models from the observational data.
- Treating missing values of the vital clinical variables, when designing the machine learning models, poses great challenges for the scientists.
- Teaching the physicians how to understand the applicability of the machine learning model's output for patient care can be a challenge.

- The results obtained in a machine learning model cannot be applicable to datasets that are even slightly different from the training dataset used in the model.
- It is difficult to compare the performance of one machine learning model with that of others, as the methodologies are dissimilar and applied to different populations with special sample distributions and features.
- The machine learning model built to analyze a particular problem cannot be used to derive generalized solutions, because of technical differences in computer equipment, EHR systems, coding definitions, and laboratory equipment, along with the dissimilarities in local medical and organizational practices.
- Machine learning models must be secured from malicious attack, whereby manipulating the dataset may produce incorrect results and dangerous mis-interpretations.

8.8 CONCLUSION AND FUTURE DIRECTIONS

Due to changes in lifestyle, diet, work pressure, and heredity, human beings are affected by many diseases nowadays. The healthcare industry needs some advanced strategies by whichto detect the possibility of diseases occurring in individual patients. Early detection of diseases like cancer will save the lives of people. Moreover, treatment should be personalized because the same type of treatment administered to all patients with the same disease will not be appropriate. Machine learning algorithms play a vital role in the healthcare industry by helping the doctors by maintaining the patient history in the form of electronic health records, by smart health monitoring in the form of wearable devices, by the early detection of dangerous diseases like cardiovascular problems, by allowing precision medicine for individuals, and clinical decision support systems. In this chapter, the above-mentioned techniques have all been discussed.

Most countries have poor data quality and scattered data systems, which makes it difficult to digitize the healthcare records. As a consequence, new techniques to handle data consolidation and digitization are necessary for appropriate processing and analysis. Furthermore, when developing a new model, care should be taken to allow the model to adapt to changes from time to time. For proper prediction of a disease, integrating the patient's data from various sources has to be carried out, which increases the complexity of the model. Researchers have been able to propose solutions for easy integration. Enhanced security protocols also need to be developed with secure infrastructure for processing patient data with high standards of privacy and in compliance with rigorous security standards.

REFERENCES

1. Anshari, Muhammad. "Redefining electronic health records (EHR) and electronic medical records (EMR) to promote patient empowerment." *IJID (International Journal on Informatics for Development)* vol. 8, no. 1, pp. 35–39, 2019.

2. Florence, F. Odekunle. "Current roles and applications of electronic health record in the healthcare system." *International Journal of Medical Research & Health Sciences* 5, no. 12, pp. 48–51, 2016.
3. https://www.technologynetworks.com/informatics/articles/healthcare-and-data-science-how-ehr-and-ai-can-go-hand-in-hand-332840, Accessed on June 10, 2020.
4. https://digitalguardian.com/blog/what-hitech-compliance-understanding-and-meeting-hitech-requirements, Accessed on June 10, 2020.
5. Luz, Christian F., Marcus Vollmer, Johan Decruyenaere, Maarten W. Nijsten, Corinna Glasner, and Bhanu Sinha. "Machine learning in infection management using routine electronic health records: tools, techniques, and reporting of future technologies." *Clinical Microbiology and Infection* 26(10), 2020.
6. Ye, Chengyin, Jinmei Li, Shiying Hao, Modi Liu, Hua Jin, Zheng Le, Minjie Xia et al. "Identification of elders at higher risk for fall with statewide electronic health records and a machine learning algorithm." *International Journal of Medical Informatics*, pp. 104105, 2020.
7. Nori, Vijay S., Christopher A. Hane, William H. Crown, Rhoda Au, William J. Burke, Darshak M. Sanghavi, and Paul Bleicher. "Machine learning models to predict onset of dementia: A label learning approach." *Alzheimer's & Dementia: Translational Research & Clinical Interventions*, vol. 5, pp. 918–925, 2019.
8. Wang, Yanshan, Yiqing Zhao, Terry M. Therneau, Elizabeth J. Atkinson, Ahmad P. Tafti, Nan Zhang, Shreyasee Amin, Andrew H. Limper, Sundeep Khosla, and Hongfang Liu. "Unsupervised machine learning for the discovery of latent disease clusters and patient subgroups using electronic health records." *Journal of Biomedical Informatics*, vol. 102, pp. 103364, 2020.
9. Dhillon, Arwinder, and Ashima Singh. "Machine learning in healthcare data analysis: A survey." *Journal of Biology and Today's World*, vol. 8, no. 6, pp. 1–10, 2019.
10. Roth, Jan A., Manuel Battegay, Fabrice Juchler, Julia E. Vogt, and Andreas F. Widmer. "Introduction to machine learning in digital healthcare epidemiology." *Infection Control & Hospital Epidemiology*, vol. 39, no. 12, pp.1457–1462, 2018.
11. Carnevale, Lorenzo, Antonio Celesti, Giacomo Fiumara, Antonino Galletta, and Massimo Villari. "Investigating classification supervised learning approaches for the identification of critical patients' posts in a healthcare social network." *Applied Soft Computing*, vol 90, pp. 106155, 2020.
12. Zhang, Yichi, Molei Liu, Matey Neykov, and Tianxi Cai. "Prior Adaptive Semi-supervised Learning with Application to EHR Phenotyping." arXiv preprint arXiv: 2003, pp. 11744, 2020.
13. Lütz, Elin. "Unsupervised machine learning to detect patient subgroups in electronic health records." *Degree Project in Computer Science and Engineering*, KTH Royal Institute of Technology, Stockholm, Sweden, 2019.
14. Rejab, Fahmi Ben, Kaouther Nouira, and Abdelwahed Trabelsi. "Health monitoring systems using machine learning techniques." In *Intelligent Systems for Science and Information*, (Janusz Kacprzyk ed.) Springer, Cham, pp. 423–440, 2014.
15. Machorro-Cano, Isaac, Giner Alor-Hernândez, Mario Andrés Paredes-Valverde, Uriel Ramos-Deonati, José Luis Sánchez-Cervantes, and Lisbeth Rodríguez-Mazahua. "PISIoT: A machine learning and IoT-based smart health platform for overweight and obesity control." *Applied Sciences*, vol. 9, no. 15, pp. 3037, 2019.
16. Basu, Samik, Sinjoy Saha, Soumya Pandit, and Soma Barman. "Smart health monitoring system for temperature, blood oxygen saturation, and heart rate sensing with embedded processing and transmission using IoT platform." In *Computational Intelligence in Pattern Recognition*, (Asit Kumar Das, Janmenjoy Nayak, Bighnaraj Naik, Soumen Kumar Pati, and Danilo Pelusi, eds.) Springer, Singapore, pp. 81–91, 2020.

17. Bambal, Jyoti Chandrashekhar and Roshani B. Talmale, "Designing a Disease Prediction Model using Machine Learning." *IOSR Journal of Engineering (IOSRJEN)*, Volume-17 pp. 27–32, 2019.
18. Iyyanki, Muralikrishna, Prisilla Jayanthi, and Valli Manickam, "Machine learning for health data analytics: A few case studies of application of regression," In book: Challenges and Applications for Implementing Machine Learning in Computer Vision (pp 241-270) Publisher: IGI Global Publisher, January 2020.
19. Jamgade, Akash C. and S. D. Zade, "Disease prediction using machine learning." *International Research Journal of Engineering and Technology (IRJET)*, vol. 06, no. 05, pp. 6937–6938, May 2019.
20. Sahoo, Abhaya Kumar, Chittaranjan Pradhan, and Himansu Das, "Performance evaluation of different machine learning methods and deep-learning based convolutional neural network for health decision making," In book: Nature Inspired Computing for Data Science pp. 201–212, January 2020.
21. Bhuvaneeswari, R., P. Sudhakar, and R. P. Narmadha, "Machine learning based optimal data classification model for heart disease prediction." In *ICICI 2019: Intelligent Data Communication Technologies and Internet of Things*, pp. 485–491, November 2019.
22. Dhomse, Kanchan B, and M Mahale Kishor, "Study of machine learning algorithms for special disease prediction using principal of component analysis." In *International Conference on Global Trends in Signal Processing, Information Computing and Communication*, pp. 5–10, 2016.
23. https://towardsdatascience.com/heart-disease-prediction-73468d630cfc, Accessed on June 11, 2020.
24. https://towardsdatascience.com/predicting-presence-of-heart-diseases-using-machine-learning-36f00f3edb2c, Accessed on June 11, 2020.
25. Detrano, R., A. Janosi, and W. Steinbrunn, "International application of a new probability algorithm for the diagnosis of coronary artery disease." *American Journal of Cardiology* vol. 64, no. 5, pp. 304–310, 1989.
26. Gudadhe, M., K. Wankhade, and S. Dongre, "Decision support system for heart disease based on support vector machine and artificial neural network." In *Proceedings of International Conference on Computer and Communication Technology (ICCCT), Allahabad, India*, pp. 741–745, September 2010.
27. Kahramanli, H. and N. Allahverdi, "Design of a hybrid system for the diabetes and heart diseases." *Expert Systems with Applications*, vol. 35, no. 1–2, pp. 82–89, 2008.
28. Palaniappan, S. and R. Awang, "Intelligent heart disease prediction system using data mining techniques," In *Proceedings of IEEE/ACS International Conference on Computer Systems and Applications (AICCSA 2008), Doha, Qatar*, pp. 108–115, March-April 2008.
29. Olaniyi, E. O. and O. K. Oyedotun, "Heart diseases diagnosis using neural networks arbitration," *International Journal of Intelligent Systems and Applications*, vol. 7, no. 12, pp. 75–82, 2015.
30. Das, R., I. Turkoglu, and A. Sengur, "Effective diagnosis of heart disease through neural networks ensembles." *Expert Systems with Applications*, vol. 36, no. 4, pp. 7675–7680, 2009.
31. Haq, Amin Ul, Jian Ping Li, Muhammad Hammad Memon, Shah Nazir, and Ruinan Sun, "A Hybrid Intelligent System Framework for the Prediction of Heart Disease Using Machine Learning Algorithms, Hindawi." *Mobile Information Systems*, vol. 2018, pp. 21, 2018.
32. Latha, C. Beulah Christalin, S. Carolin Jeeva, "Improving the accuracy of prediction of heart disease risk based on ensemble classification techniques." *Informatics in Medicine Unlocked* vol. 16, pp.100203, 2019.

33. https://ghr.nlm.nih.gov/primer/precisionmedicine/definition, Accessed on June 12, 2020.
34. Emmert-Streib, Frank and Matthias Dehmer, "A machine learning perspective on personalized medicine: An automized, comprehensive knowledge base with ontology for pattern recognition." *Machine Learning & Knowledge Extraction*, vol. 1, pp. 149–156, 2019.
35. Uddin, Mohammed, Yujiang Wang and Marc Woodbury-Smith, "Artificial intelligence for precision medicine in neurodevelopmental disorders." *npj Digital Medicine*, vol. 2, pp. 112, 2019.
36. https://www.forbes.com/sites/insights-intelai/2019/02/11/how-machine-learning-is-crafting-precision-medicine/#46e8abf35941, Accessed on June 12, 2020.
37. https://emerj.com/ai-sector-overviews/machine-learning-in-pharma-medicine, Accessed on June 12, 2020.
38. Sutton, Reed T., David Pincock, Daniel C. Baumgart, Daniel C. Sadowski, Richard N. Fedorak and Karen I. Kroeker, "An overview of clinical decision support systems: benefits, risks, and strategies for success." *npj Digital Medicine*, vol. 3, no. 17, 2020.
39. Zikos, Dimitrios and Nailya DeLellis, "CDSS-RM: a clinical decision support system reference model." *Medical Research Methodology*, vol.18, pp. 137, 2018.
40. Huang, M., H. Han, H. Wang, L. Li, Y. Zhang and U. A. Bhatti, "A clinical decision support framework for heterogeneous data sources." *IEEE Journal of Biomedical and Health Informatics*, vol. 22, no. 6, pp. 1824–1833, November 2018. doi: 10.1109/JBHI.2018.2846626.
41. Buchlak, Quinlan D., Nazanin Esmaili, Jean-Christophe Leveque, Farrokh Farrokhi, Christine Bennett, Massimo Piccardi, and Rajiv K. Sethi, *Machine Learning Applications to Clinical Decision Support in Neurosurgery: An Artificial Intelligence Augmented Systematic Review, August 2019.* Springer-Verlag GmbH Germany, part of Springer Nature, 2019.
42. Mazo, Claudia, Cathriona Kearns, Catherine Mooney, and William M. Gallagher, "Clinical decision support systems in breast cancer: A systematic review." *Cancers (Basel)*, vol. 12, no. 2, pp. 369, February 2020.
43. Vitabile, Salvatore, Michal Marks, Dragan Stojanovic, Sabri Pllana, Jose M. Molina, Mateusz Krzyszton, Andrzej Sikora et al. "Medical data processing and analysis for remote health and activities monitoring." In *High-Performance Modelling and Simulation for Big Data Applications*, (Joanna Kolodziej, Horacio Gonzalez-Velez, eds.) Springer, Cham, pp. 186–220. 2019.
44. Jiang, Yicheng, Bensheng Qiu, Chunsheng Xu, and Chuanfu Li, "The research of clinical decision support system based on three-layer knowledge base model." *Hindawi Journal of Healthcare Engineering*, vol. 2017, Article ID 6535286, 2017.

9 Applications of Computational Methods and Modeling in Drug Delivery

Rishabha Malviya and Akanksha Sharma

CONTENTS

9.1 INTRODUCTION

Scientists can now employ technology in computer development and modeling tools, and can use it for improved understanding and visualization of a new system of drug delivery. Computer modeling for the delivery of drugs is an exciting area where interdisciplinary collaborations address very complex delivery of drug challenges. This multidisciplinary method involves engineers, mathematicians, and physicians. For many years, polymeric systems were used in pharmaceuticals to ensure controlled release of therapeutic drugs. Some polymer and drug systems may be useful in shielding drugs from biological degradation in the body before they are released.

These novel approaches to the drug delivery system involve computer-assisted drug formulation design (CADFD), which plays an important role by using various informatics methods, like molecular docking, molecular dynamics, artificial neural networks, and data mining to estimate the different drug loads on their potential carriers. In 2015, Metwally *et al.* used computational informatics and tools for estimating the loading of drugs on a specific carrier. They effectively demonstrated that artificial neural networks (ANN) can describe strong relationships on highly predictive carrier systems between docker binding energies and certain molecular drug descriptors.

Quantum mechanics defines the highest level of the system. Quantum mechanical calculations are commonly applied to carbon nanotubes (CNTs), which play a significant role in the optimization of less complex simulations, but they are used only in small systems with minimal degrees of freedom. Molecular dynamics (MD) is the technique that is generally used for soft-condensed and biological systems. This approach is used in the majority of cases, and is discussed in various papers. MD is a method used commonly in many software packages, including GROningen Machine for Chemical Simulations (GROMACS), Assisted Model Building with Energy Refinement (AMBER), Chemistry at Harvard Macromolecular Mechanics (CHARMM), Nanoscale Molecular Dynamics (NAMD), and Large-scale Atomic/Molecular Massively Parallel Simulator (LAMPPS). Another technique used is Coarse-grained (CG) which is frequently based on the MARTINI model [1].

In the past few years, *in vitro* drug release from controlled-release platforms has been combined with state-of-the-art modeling, i.e. Computational Fluid Dynamics (CFD) simulation, which helps to predict the temporal and spatial variation of drug transport in living tissues. The simulation is conducted on CFD tools to explain drug momentum, continuity, and concentration equations simultaneously. The function of macromolecular and micromolecular convective transport in the tumor vicinity has also been studied.

Quantitative modeling of biological phenomena includes techniques used by other quantitative fields, like mathematics, informatics, physics, and engineering, which are used to deal with the biological complexity. Biological systems have vast complexity and quantitative evaluation parameters, which are generally difficult to obtain, with the results often depending on the modeling assumptions. Other approaches to computational dynamics include Boolean networks, Petri nets, agent-based models, cellular automata, Bayesian network statistical methods, and formal systems like algebraic processes.

There are four distinct modeling methods that describe the delivery of a drug from the central blood vessel to the tumor cord. These models are the binding model, the multi-dimensional cell-center model, the radially symmetric compartment model, and the radially symmetric continuum model.

The theoretical approach to computerized modeling indicates that the business model is a computerized representation of the structure, operations, procedures, knowledge, resources, workers, attitudes, objectives, and constraints of an organization. Thus, from the design stage point of view, the enterprise model must offer as explicitly as possible a specialized language intended to define the enterprise. In general, the classical models of the life cycle of the software products include four major stages: preparation and selection, review, design, and implementation and maintenance, to which additional stages may be added if the situation requires.

In the past decade, Computer-Aided Drug Delivery (CADD) has re-emerged to reduce the number of the compounds required for screening, while maintaining the same success rate for the discovery of lead compounds. Several compounds that are predicted to be inactive may be skipped and priority may be given to those predicted to be active. It reduces the costs and workload of a full High-Throughput Screening (HTS) program, without sacrificing lead discoveries. However, conventional HTS screens often need comprehensive validation and development before they are used. CADD requires considerably less time in terms of preparation than conventional screening, and experiments are conducted during the traditional HTS assay preparation. An additional advantage of CADD in the discovery of drugs is that these methods are used in parallel [9]. For methods complementary to experiments, computer modeling mainly sheds light on the dynamic and structural properties of structures at the molecular or atomic levels. Computer simulations related to Drug Delivery Systems (DDS) are used to answer a wide variety of queries. Computer simulations are used to study drug charge capacity, self-assembly, and dynamic and structural characteristics of the resulting rate and mechanism, aggregates, distribution of the drug or DDS localization, drug retention, complex stability, release rate, and release mechanism, to optimize or design DDS with targeting capabilities and dominant interactions. Environmental factors, like temperature, pH, counterions, salt form and concentration, and external stimuli, such as external magnetic fields and interactions with other biomolecules, such as microRNA (miRNA), heparin, and serum proteins, may all influence the above-mentioned aspects of DDS and may be studied in a

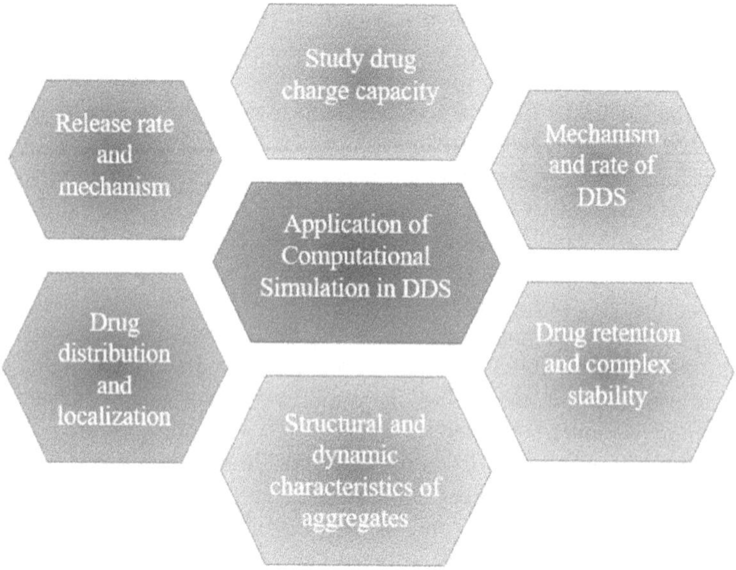

FIGURE 9.1 Schematic diagram of the use of computational simulation in DDS.

quantitative manner [2]. Figure 9.1 shows the application of computational simulation in DDS.

9.2 COMPUTER-AIDED DESIGN FOR FORMULATION

CADD is the combination of information on the macromolecular structures of biochemicals, their properties, and interactions. This information is turned into insights that support research to make better drug and production decisions. The relevant, comprehensive, and ever-increasing data must be investigated by using supporting computational and chemical tools, as well as bioinformatics. CADD includes computer data storage, management, analysis, and software tools and resources used in drug modeling. This involves the development of digital tools to analyze chemistry. Computer programs are used to design compounds that have interesting physicochemical properties, with identification methods which help in the systematic evaluation of lead compounds before they are synthesized and evaluated. In the early 1970s, the foundations of CADD were established by their applications to structural biology, such as in altering the biological activity of insulin and directing the synthesis of human hemoglobin ligands.

Computing methods are attracting considerable research, implementation, and admiration as part of drug design, research, and production processes. The introduction of a new drug onto the market is a complex, dangerous, time-consuming, and costly process. It is usually found to take between 10 and 14 years to discover, assess, and market the drug, with a total capital of more than one billion US dollars per drug. Therefore, CADD methods are commonly employed as a modern approach

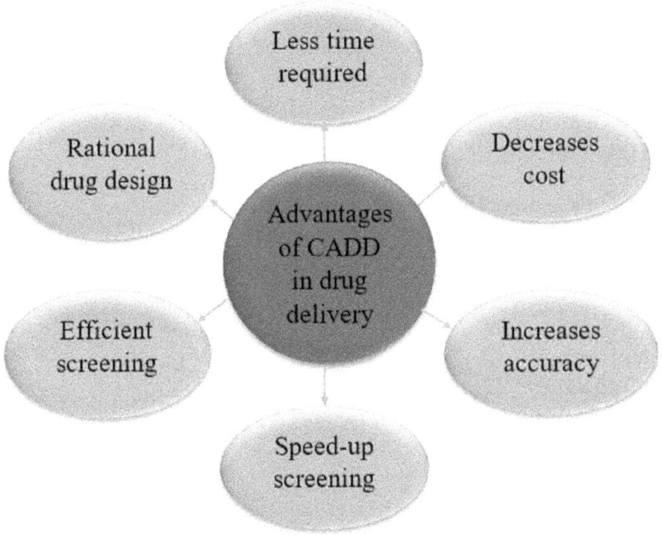

FIGURE 9.2 Schematic diagram of the advantages of CADD in drug delivery.

to design the drug formulation, reducing the time, cost, and risk. It has been seen that the expense of drug discovery and production can be halved through the use of CADD approaches. For CADD to set a standard for a structure-activity relationship, any programming process and software can be used

A new computational approach for calculating the load of individual drugs on their respective potential carriers is known as computer-assisted drug formulation design (CADFD). When predicting the respective energy values from major physicochemical and electronic descriptors, the loading results are linked with the amortizing energies by using artificial intelligence technologies, like artificial neural networks [3]. The advantages of CADD in drug delivery are shown in Figure 9.2.

9.2.1 Advantages of CADD

Advantages of CADD include:

- CADD decreases the effort associated with synthetic and biological research.
- CADD identifies the most suitable candidate compound by eliminating compounds with inappropriate properties (low effectiveness, weak absorption, distribution, metabolism, excretion, or toxicity (ADMET), etc.). from *in silico* filters.
- CADD is a time-saving, quick, streamlined process, which make it cost effective.
- CADD provides insights about the sequence of interactions between drugs and receptors.

- CADD offers high-hit compounds by searching large *in silico* compound libraries, compared with traditional high-performance screening processes.
- CADD strategies minimize the probability of final phase failure [4].

9.2.2 CADD Approaches

CADD approaches used in the development of drugs are of two types:

1. Structure-based drug design, or the "direct approach".
2. Ligand-based drug design, or the "indirect approach".

9.2.2.1 Structure-Based Drug Design (SBDD)

SBDD is employed to recognize target protein structures and it also measures an association, or bio-affinity, of a novel drug molecule for improved interaction with the target protein.

The SBDD process investigates the details of the three-dimensional structure of macromolecular targets, such as RNA or protein, to recognize main interaction sites relevant to their biological functions. The Protein Data Bank (PDB) is used to determine the 3D structures of the protein, RNA or other target macromolecules, as identified by experiments involving nuclear magnetic resonance (NMR) or X-ray crystallography. Alternatively, by using homology modeling methods, a 3D structure can be developed with a program like MODELLER, or an online web server, like SWISS-MODEL [5]. Structure-based drug design approaches are described below.

9.2.2.1.1 Structure-Based Virtual High-throughput Screening

Structure-based virtual high-throughput screening (SB-vHTS) is an *in-silico* technique used to determine the putative hits to targets on the known structure of various compounds, typically focused on the docking of molecules. In the protein active site models, a small molecule is incorporated and a comparison is carried out between the small molecule's 3D structure and the putative binding pocket. In conventional HTS, the ligands can generally attach to, modify, or inhibit the protein, whereas, in SB-vHTS, the ligands are selected which are required to adhere to a particular binding site. To confirm the feasibility of large numbers of compounds over a finite time, SB-vHTS employs protein-selective conformational sampling and ligands, along with energy binding measured [6].

9.2.2.1.2 Structure-Based Virtual Screening

Structure-based virtual screening (SBVS) is a technique employed in the initial stage of drug discovery. It scans new bioactive molecules against a certain target site of the drug from a library of chemical compounds. The significance of screening includes increased hit rate by reducing the number of the compounds which are assayed during research for their behavior, and thus increasing the success rate of the *in-vitro* experiments. This approach is widely used in the early stage of drug development by pharmaceutical companies and research groups [7].

9.2.2.1.3 Fragment-Based Lead Compound Discovery

The first fragment-based discovery report was published in 1996, and, by the late 1990s, the methods and their implementation had rapidly evolved. In the early 2000s, methods specifically aimed at small molecule structure-based discovery were performed by companies like Sunesis, Vernalis, and Astex. Now, the fragment methods have been established for most 'conventional' drug discovery to target specific sites, such as enzymes (principally proteases and kinases), but also for more complex targets, like protein–protein interactions, where disruption can also be increased. The main application is to identify tractable chemical starting points that modulate the behavior of a biological molecule non-covalently. The main feature of fragment methods is that discovery starts with the detection of compounds or fragments of low molecular weight (typically about 200 Da), which bind with the target molecule of interest. Fragments bind to maximum binding sites of the target with an equilibrium dissociation constant (KD_a) range between 100 μM to low mM. This requires tests that can reliably detect fragment design (high solubility needed in assays for high fragment concentrations) and weak binding (using biophysical methods) [8].

9.2.2.1.4 In-Silico *Structure-Based Lead Compound Optimization*

This approach increases the lead compound optimization by describing its properties relative to pharmacology, after which the desired compounds are detected through virtual screening, thereby reducing the experimental time of *in-vivo* and *in-vitro* studies [6].

9.2.2.1.5 ADMET Modeling

Efficient and safe drugs show a balanced combination of pharmacokinetics (PK) and pharmacodynamics (PD), including high affinity, potency, and selectivity toward the molecular target, with sufficient ADMET. There is a wide range of ADMET prediction tools available today, such as DataWarrior, QikProp, MetaTox, StarDrop, and MetaSite. Because of the safety issues and loss of effectiveness associated with drugs, the research and development field plays an important part in the attrition rates of drugs, which influences the PK properties, which have declined over the past decade. This reduction comes from the improved PK control systems in the research pipeline and their increasingly earlier implementation. Completely integrated ADMET prediction systems can easily remove inappropriate compounds by targeting several PK parameters simultaneously, decreasing the synthesis evaluation cycles, and eliminating the more costly failures which occur at a late stage [9].

9.2.2.2 Ligand-Based Drug Design

The target protein 3D structure is not identified by ligand-based drug design (LBDD) but the ligands which bind to the anticipated site of the target molecule are identified by LBDD. Such ligands are used for the creation of a pharmacophore model or molecule, that has all the structural features required to link to the target active site. A pharmacophore-based approach and quantitative structure–activity relationships (QSAR) are typically ligand-based techniques. In LBDD, the protein is considered to be similar in structure when the compounds have similar biological activities and interactions with the target [10].

These approaches are used where the spatial structure of the targeted macromolecule is uncertain and an accurate model of it cannot be built. These approaches are focused on the analysis of ligand sets with their biological activity. These include the pharmacophore model design, the study of the relationship between quantitative structure, behavior and its 3D-QSAR (Quality Structure Activity Relationship) adjustment, the quantitative structure–properties relationship (QSPR), etc. The LBDD approaches are used for lead compounds to discover and optimize the identified ligands. Pharmacophore models or separate active site cavity models are used to this end. The pharmacophore model characterizes a series of space points with specific properties and distances between them, which helps to determine the binding of a given ligand group with the target. Pharmacophore points can be the cyclic groups, atoms, acceptor, or donor atoms that are charged positively or negatively. Pharmacophore models based on molecular database mining enable the selection of molecules that satisfy this model. However, the identification of new classes of chemical compounds by this route is almost impossible; typically, the compounds chosen represent analogs of known structures. Another molecular database mining approach uses the compounds docking into the spatial structure model of the enzyme's active site. These approaches are based upon an interpretation of recognized ligand structures and properties [11].

The ligand-based computer-aided drug discovery method (LB-CADD) includes ligand evaluation, which interacts with an interesting target. These approaches employ known compound structures to interact with the subject of interest and examine their 3D and 2D structures. The ultimate aim, when designing these compounds, is to preserve the most significant physicochemical properties for the preferred interactions, while, at the same time, discarding the properties which are not essential for the interactions. This is called an indirect approach for the discovery of a drug. The two basic approaches of LB-CADD are (a) the selection of the compound is established on the grounds of chemical resemblance to known active ingredients, having a few degrees of similarity, or (b) the development of a QSAR model which predicts the chemical structures with appropriate biological activity. With *in-silico* screening, the methods are applied to new compounds with interesting biological activity, leading to a hit-to-lead drug optimization, as well as to Drug Metabolism and Pharmacokinetics (DMPK)/ADMET property optimization [12]. Ligand-based modeling is used in various approaches which are described below.

9.2.2.2.1 QSAR

Three-dimensional QSAR (3D-QSAR) is an important modification of QSAR, in which the chemical characteristics of active ligands are accumulated to create a particular model in 3D space to target the active sites. The QSAR studies are based on the assumption that changes in the bioactivity of compounds is correlated with molecular and structural variations. A statistical model for the creation and mathematical prediction of the biological properties of novel compounds is developed from this correlation. The development of a consistent QSAR model involves several restrictions: bioactivity data must be of an adequate number (a minimum of 20 active compounds) and obtained from a standard protocol of experimentation to make the potency values comparable; proper selection of compounds to test a training set; non-autocorrelation of ligands as molecular descriptors to prevent overfitting; and validation of the model, using external and internal validation to evaluate its predictivity and applicability. Comparative molecular field analysis (CoMFA), developed over three decades ago, continues to be the method most commonly used with the 3D-QSAR model. Currently, 3D-QSAR techniques include CoMFATopomer, Spectral Structure Behavior Relationship (S-SAR), Molecular Comparison Field Adaptation (AFMoC), and Residue Interaction Comparative Analysis. Despite its notable drug discovery achievements, 3D-QSAR has a number of inadequacies that are solved by using extra advanced multidimensional QSAR methods, such as 4D, 6D and 5D-QSAR. The method 4D-QSAR is designed to resolve the conformation of the ligand and the site of target binding, whereas 5D-QSAR addresses problems, such as receptor versatility and incorporated fit effects. The method 6D-QSAR shows the solvation effects which integrate its important function in receptor–ligand interactions. Developments in computing software and power efficiency were also used by the Discovery Bus and AutoQSAR to improve the creation and validation of QSAR models. The statistically highly predictive models are modified, discovered, and tested critically in both approaches, by constantly incorporating new agents of machine learning and descriptors into the framework [13].

9.2.2.2.2 Pharmacophore Modeling

A target binding site pharmacophore model precisely describes the electronic and steric features essential for a ligand which optimally interacts with the target. Hydrogen bond acceptors, acid groups, hydrogen bond donors, aliphatic hydrophobic moieties, basic groups, partial charges, and aromatic hydrophobic groups are the most common properties, which are used to describe pharmacophores. Pharmacophore technologies have been commonly used for virtual screening, and the lead optimization and the *de novo* design model are used in drug discovery. The target binding site pharmacophore model is used in practice to scan a library of the compound for putative hits. In addition to querying the database for active compounds, *de novo* design algorithms can also utilize the pharmacophore models to direct new compound developments [14].

9.2.2.2.3 Structure-Based Pharmacophore Modeling

Structure-based pharmacophore methods are built based on either a target binding site analysis or on the complex target ligand structure. Ligand Scout comprises the complex data of the protein–ligand associations between destination and ligand. An information-based rule collection from the PDB is utilized to define and classify interactions into hydrogen bonds, lipophilic regions, or transfer of charges. The Pocket v.2 algorithm is capable of constructing a target–ligand complex by using an automated pharmacophore model. The algorithm creates the normally spaced grids around the residues surrounding the ligand. The grids are scanned with sample atoms representing a hydrogen bond acceptor, hydrogen bond donor, or a hydrophobic group. To define a constant binding between the probe atoms and the target, an analytical scoring method, the score, is used. The score comprises terminology to determine the interactions between hydrogen bonding, van der Waals forces, desolvation binding, and metal–ligand bonding effects [15].

9.2.2.2.4 Pharmacophore Model-Based Virtual Screening

The ligand-based approach or the structure-based approach produces a pharmacophore model. It is used to query the database of 3D chemical structures, which is scanned for possible ligands, to result in pharmacophore-based virtual screening (VS). At present, docking-dependent VS and pharmacophore-dependent VS represent the majority of VS methods. Unlike VS methods counterpart, the pharmacophore-based VS and docking-based VS processes decrease as a result of insufficient protein flexibility and the use of inadequately optimized or designed scoring functions, by integrating a tolerance radius and pharmacophoric functionality. A pharmacophore hypothesis is used as the basis of the pharmacophore-based approach to the VS. The screening aims to identify the molecules which have chemical characteristics similar to those of the templates. The hunt for compounds with different scaffolds but similar biological functionality is typically called 'scaffold hopping' [16].

9.3 MOLECULAR DYNAMICS

The science which simulates the time-dependent behavior of a group of particles is known as molecular dynamics. The basic elements for simulation of molecular dynamics are the potential for interaction, such as the potential energy of the particles, from which the forces can be measured, and Newton's Laws of Motion, governing the particle dynamics. At each stage of the molecular dynamic simulation, the atom's position and its speed are modified. The field of force is a collection of physical constants and equations, which is used to measure these forces. Although molecular dynamics are commonly used in biophysical research, fewer studies have explored their use in the formulation of drugs [17].

The simulation of molecular dynamics (MD) is one of the main methods for simulating biological molecules theoretically. Because of the large number of particles in molecular systems, these complex structures cannot be analyzed. The numerical methods are used to prevent the intractability of the MD simulation. Molecules and atoms can interact for a certain time during the simulation. The movement of each atom is measured and the resulting behavior can be studied. Ultimately, the algorithm for MD simulation includes determination of the speed and initial positions of each atom and determination of the force, which is used in the analysis of the atoms, by using inter-atomic potential and advancing the speed and atomic position over a short period of time [18].

Simulations of MD are valuable instruments for structural drug design. In the high-throughput docking campaigns, initial or final stage MD was used. In a few instances, protein-targetted MD simulations are performed before docking to generate a protein conformer, that varies from existing crystal structures. Also, MD runs performed after docking are used to evaluate the top-ranking compound binding modes or to guide the chemical synthesis to optimize the hit [19].

MD simulations analyze the space of configuration and generate a trajectory that acts as a function of time, following molecular movements. MD simulations are now regularly applied to pharmaceutically and biologically engineered macromolecular systems, initially designed for the study of liquid state properties. Applications involve refinement of experimentally defined structures, homology modeling and conformational analysis of proteins, interpretation of atomic-level biophysical and biochemical processes, docking of biomolecular complexes, and several methods for measuring free energy adjustments [20].

9.4 MOLECULAR DOCKING

Molecular docking is a method of molecular modeling that predicts whether, if connected, one molecule is preferred to orientate in a stable complex [36]. The chosen orientation is used to determine the two molecules scoring functions whether to have a strength or binding affinity. Interactions between biologically essential molecules, like proteins, carbohydrates, lipids, and nucleic acids, play a major role in the transduction of signals. The relative orientation can also affect the produced signals (agonism). Hence, docking is used in the prediction of both the power and the signal type which is produced. To measure affinity and the activity of the small molecules, docking is also used to prevent drug candidates from interacting with their protein targets. Therefore, docking performs a significant role in the logical design of medicines.

The goal of molecular docking is to attain optimized protein and ligand conformations, and to decrease the overall energy of the system by making a relative re-orientation between ligands and proteins. The promotion of biomolecular fundamental activities, such as drug-proteins, enzymesubstrates, and interactions of the drug with a nucleic acid, plays a key role in molecular recognition. Molecular docking-based virtual screening is becoming a major part of various structure-based processes of drug discovery. Therefore, evaluating current docking schemes is now an important task, helping to decide on the best algorithm for docking for each analysis. The enzymes become inhibited or activated by binding of a small ligand molecule to an enzymatic protein. If the protein is a receptor, binding can induce antagonism or agonism [21].

9.4.1 Application of Docking

Applications of docking can be categorized into three classes, namely hit identification, lead compound optimization, and bioremediation protein.

9.4.1.1 Hit Identification

Using the docking feature, a large database of drugs with desirable features may be rapidly screened *in silico* and the molecules which bind to their selected protein target can be identified (virtual screening) [22].

9.4.1.2 Lead Optimization

Through docking studies, a ligand binding with the target protein (site and type of orientation) can be anticipated to provide relative guidance, that can lead to the design of analogs with greater power and selectivity [23].

9.4.1.3 Bioremediation Protein

The docking of protein ligands may also be used to avoid the degradation of toxins by enzymes. The binding affinity is calculated [24].

9.5 ADVANCES IN DEEP LEARNING APPROACHES

The deep learning system is mainly associated with the artificial neural network (ANN) and the convolutional neural network (CNN), which are used in the layering concept of learning. There are many approaches to learning, from low to high levels. The deep learning approach automatically chooses representations from high-dimensional data and the original data if the molecular descriptors are not selected.

9.5.1 Artificial Neural Network

Black box modeling algorithms, like ANN, model nonlinear systems and, as a result of their accuracy, simplicity and high predictive efficiency, have gained widespread acceptance from the scientific community. ANN are mathematical models that can learn the relationship between response (output) values and experimental (input) values through error testing and iterative testing. This is the form of artificial intelligence that simulates the learning of the human brain. Data processing and pattern recognition are usually carried out even in the event of complex multidimensional problems. In one analysis, the microemulsion form was forecast from the composition or its Differential Scanning Calorimetry (DSC) curves by two evolutionary ANN. Both ANN reported predictive accuracies of greater than 90 percent [25].

The ANN system is based on computer technology that allows for the simulation of the human brain's neurological processing capacity. ANN has been used successfully to address different issues in pharmaceutical research, including product development, the calculation of the diffusion coefficient, and prediction of drug function and pharmacokinetic parameters. For example, working on cellulose polymers containing a hydrophilic matrix capsule system, Hussain and colleagues [26] employed ANN to examine quantitative relationships between different formulating factors and the release parameters. ANN forecasts were found to be more accurate than the polynomial equations had predicted [26].

ANN offers a virtual tool for the estimation of the drug impact on the body, for the eventual purpose of enhancement of patient care by pharmaceutical development. A dynamic and interdisciplinary field, it involves computational biologists and pharmacologists in the creation of the ANN for drug disposition. ANN reduces the demand for conventional pharmacological testing, including clinical trials, thereby minimizing the time and cost of drug development. The software is not only used to create novel drugs but also to enhance the quality of existing products. The accuracy of ANN in predicting routes for drug excretion is still a work in progress. The 'system pharmacology' has been opened up by using biological data obtained from the post-genomic era. There is a wide range of biological datasets in pharmaceutical systems, that provide a broad range of information, including proteomics, genomics, drug–protein interactions, gene expression, and metabolomics. These datasets can be used as ANN inputs to react to different types of questions related to pharmacology.

ANN can predict the various properties of chemical compounds. ANN is a valuable tool for preformulation studies, predicting the physicochemical characteristics of amorphous polymers. The model is capable of accurately visualizing the relationships between the chemical composition of the polymers and the absorption profiles of water, the transition temperature and viscosity of glass. These results show the capability of the ANN model in preformulation studies. The nonlinear pattern recognition capabilities of ANN models can also be used in optimizing medication formulations and dosing methods. It has been shown that ANN is capable of predicting the bioavailability profiles and *in-vivo* dissolution time profiles of formulations, enabling the identification of formulations with certain desired properties. Independent variables of the ANN model, such as the temperature of the solvent, the surfactant concentration, and the solvent and antisolvent rates of flow, are used to optimize the polydispersity index of acetaminophen nanosuspensions. This model was used to determine the flow rates of high antisolvent, low solvent flow rates, temperatures of the solvent, and low polydispersity index (PDI) [27]. In Figure 9.3, **the various applications of ANN in the delivery of drugs and other research areas are shown.**

The various ANN applications in drug delivery and research are described in detail below.

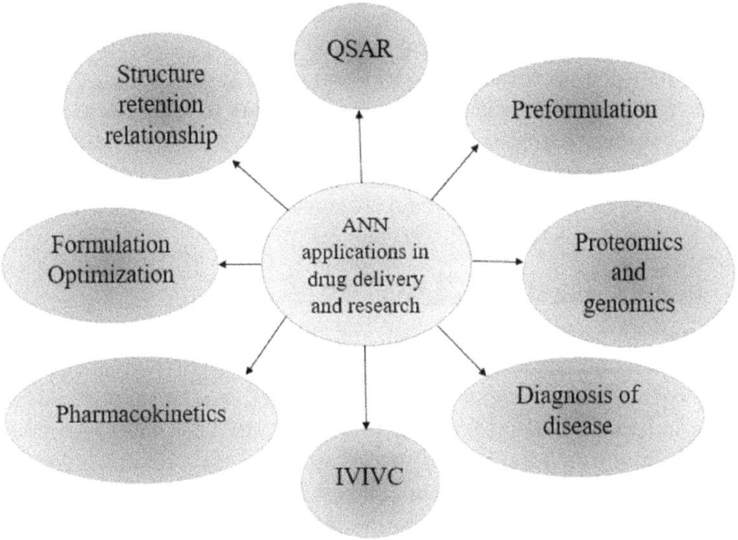

FIGURE 9.3 Schematic diagram of ANN applications to the delivery of drugs and research.

9.5.1.1 Preformulation

The ANN model was used to investigate the physicochemical properties of amorphous polymers in preformulation design and optimization. An analysis by Ebube *et al.* [28] suggested that, with a low prediction error, the ANN model correctly forecast the water absorption, glass transition temperatures, and the viscosities of various hydrophilic polymers with their different physical blends. The ANN algorithm models the preformulation study because the qualified ANN model allows prediction of the relationships between water absorption profiles and polymer mixture compositions, between polymer solution viscosity and polymer mixture composition, and between glass transition temperatures and polymer moisture content. In the design of this phase, it is necessary to determine the effect of the active pharmaceutical ingredients (API) on the tablet, particularly when different APIs are present [28]. Onuki and coworkers [29] used ANN to study 14 APIs after formulation base optimization, using self-organizing maps to extend the tablet database for fast release and investigating the impact of API on properties of the tablet. ANN has also been used widely in the design of stable formulations of many active ingredients, like isoniazid and rifampicin microemulsions. ANN is mainly used in the design of preformulations, which can help to minimize the cost and time of preformulation analysis [29,30].

9.5.1.2 ANN for Structure Retention Relationship

Predicting chromatographic activity from the solute molecular structure is the key objective of the technique of structure retention relationships. The use of ANN for modeling retention times in High Performance Liquid Chromatography (HPLC) optimization compares the chromatographic activity (capacity factors) of the solute with the pH and composition of the mobile phase. HPLC retention data have been used over the past few years as a pseudomolecular descriptor to determine the aqueous solubility of organic non-electrolytes and aromatic hydrocarbons, to evaluate the partition coefficient of octanol and water, to assess the log P_{oct} values of compounds related to chlorosubstituted aromatics, and to reliably estimate the pK_a dissociation constant. Methods of computer simulation have been used to determine the separation of compounds on the basis of pH and solvent intensity change for high-performance reversed-phase liquid chromatography, and coefficients of hydrophobicity for peptide elution profile estimations [31].

9.5.1.3 Pharmaceutical Formulation Optimization

In the optimization of the formulation, the pharmaceutical response estimation, based on the response surface methodology (RSM) and polynomial equations, are commonly used but are inadequate for low volumes, leading to poor estimates of optimal formulations. To decrease these limitations, an ANN was developed, which offered a multi-objective technique of optimization. The optimization was performed for theophylline release from controlled release tablets under appropriate drug release profiles. For ANN application, 16 drug release profiles of different formulations were observed experimentally. Optimization is achieved by decreasing the generalized difference between every estimated response value, with the desired one being optimized individually [32]. Peh *et al.* [33] conducted a similar study with optimization of the tablets for glyceryl monostearate matrix. A multilayer perceptron (MLP) neural network was trained and the similarity factor (f_2) was used to compare the expected and the actual profiles of reference dissolution. The effect of formulation optimization on drug release profile, as well as on drug stability prediction, was demonstrated by the ability of the generalised regression neural network (GRNN) to formulate controlled release preparations. This type of network appears to have the potential to optimize the managed release of formulations better than the MLP network. GRNN training is also much quicker than for other forms of networks [32,33].

9.5.1.4 In-Vitro/In-Vivo Correlations

An effective *in-vitro/in-vivo* correlation (IVIVC) model is a technique for prediction of the *in vivo* bioavailability of particular drugs based on *in vitro* results and it is easily obtained by using inexpensive and reproducible dissolution tests. This correlation enables the evaluation of bioperformance of different dosage formulations without using animal or human experiments. Neural engineering in pharmaceutical science is already well known. ANN has been successfully applied in pharmaceutical technology to optimize formulations and to optimize the preparation process, with different applications of ANN in pharmacokinetics being demonstrated [34]. Furthermore, IVIVC has also been studied in neural modeling. IVIVC applications use these ANN networks as a nonlinear *in-vitro* and *in-vivo* profile mapping method but have been restricted to single drugs and formulations [34].

9.5.1.5 ANN in Quality Structure–Activity Relationships

The Free-Wilson, and Hansch-Fujita models introduced the traditional 2D-QSAR methods. These are based on the presence or absence of physicochemical properties. In addition to the 2D QSAR methods, CoMFA or 3D QSAR, 5D QSAR, 4D QSAR and 6D QSAR were also developed. All methods, however, need manual intervention and superimposition. Partial least and multiple linear square regressions are the basis of conventional QSAR methods and can interpret linear relationships, but ANN has the added benefit of solving nonlinear relationships. ANN versatility helps to reveal more complicated relationships. ANN is used to predict biological activities of unknown compounds Multilayered feed-forward networks can be used by employing different connectivities, such as quantum, constitutional, electrostatic, topological, or the Kier Hall descriptors, to determine the solute–solvent interactions. Because of the accuracy and speed at which information processing takes place, the use of ANN in QSAR studies has increased the speed of analysis, as a result of which almost 35 different types of ANN are utilized in the development of drugs. Myint *et al.* investigated a novel ANN-based QSAR model, which trains the ANN by using molecular fingerprints. The authors predicted the binding affinities of a cannabinoid ligand for the G-protein-coupled receptor (GPCR), using FANN-QSAR 2D fingerprints [35].

9.5.1.6 ANN in Proteomics and Genomics

There has been a substantial increase in the applications of proteomic and genomic technologies recently, resulting in complex data and a high-dimensional explositation This improves the efforts of the bioinformatics community to invent new computational methods that would allow the extraction of meaningful information. Bala *et al.* [36] described the main use of ANN to be for investigation of mass spectrometry (MS) data in the astrocytoma classification. They demonstrated the initial promise of using Surface Enhanced Laser Desorption/Ionization-Time of Flight (SELDI-TOF) MS tool in cancer patients, combined with intensive protein expression screening computer algorithms [36]. Rogers *et al.* [37] also utilized the SELDI-TOF MS in an investigation of urinary proteins associated with renal cancer. ANN was used to detect early disease onset and to identify indicative biomarkers [37]. Chen *et al.*, [38] aware of the absence of biomarkers for colorectal cancer, suggested the use of proteomics, combined with analysis by ANN, to discover proteins capable of distinguishing patients from a healthy population from colorectal cancer patients [38].

Khan *et al.* [39] used principal component analysis (PCA) with ANN to classify 88 round blue-cell tumors into four diagnostic groups of more than 6,000 genes, based on complementary DNA microarray (cDNA MA) analysis. The authors achieved the ANN-based approach, which is an alternative to routine histological diagnosis, due to the high precision of the models. This dataset has been made accessible to the scientific community for free download and provides the basis for many more experiments, using different algorithms of ANN, to classify these samples effectively [39].

9.5.1.7 ANN in Pharmacokinetics

In pharmacology, ANN has been found to have broad applications, including Pharmacokinetics and Pharmacodynamics tracking, multiple PK interactions in biological systems, and prediction of PD parameters, like protein-bound fractions, rates of clearance, and volume distribution. ANN was used by Hussain *et al.* for the PK parameter prediction which is investigated in animal studies [26]. The back-propagation of artificial neural networks (BP-ANN) has been used on losartan pharmacokinetics in the rabbit to create a pharmacokinetic model. The BP-ANN model has exceptionally accurate predictive capabilities, that are higher than the curve estimate and can be used as a tool in pharmacokinetic experiments. The results disclosed that the BP-ANN model had efficacy in the high-fit index and great consistency ($r > 0.99$) where r is correlation coefficient following both intravenous and intragastrics administration between expected concentration and measured concentrations. The concentrations estimated by the BP-ANN model were lower than the curve estimates [40].

9.5.1.8 ANN in the Permeability of Skin and Blood Brain Barrier

In the circulatory system, there is a physical barrier, known as the blood-brain barrier (BBB) which prevents the passage of certain substances into the central nervous system (CNS). Drugs used as therapeutic agents and proposed to interact with molecular targets within the CNS must cross the BBB. Garg *et al.* [41] developed an ANN model to estimate the steady-state concentrations of drugs in the brain relative to those which were present in the blood (logBB) by using various molecular structural parameters. For model growth, seven descriptors, including p-glycoprotein (P-gp) substrate, are utilized. The model that has been developed can capture a relationship between logBB and P-gp [41].

Chen *et al.* [42] reported that the ANN model predicts the permeability of skin to chemicals by using Abraham descriptors. A broad skin permeability database was collected from the literature, containing 215 data points. Significant positive correlations were obtained between the coefficient of skin permeability and the Abraham descriptors of the qualified neural network model . Furthermore, the estimation of the neural network model was compared with the multiple linear regression (MLR) models. The ANN model provides better predictive results, indicating that the relationship between skin permeability and Abraham descriptors is non-linear and complex [42].

9.5.1.9 Diagnosis of Disease

Based on clinical chemical data, ANN is used in the diagnosis of diseases, such as cancer, lupus erythematosus early diagnosis, acute myocardial infarction diagnosis, cardiovascular risk assessment, assessment of developing hypertensive disorders caused by pregnancy, diagnosis of Alzheimer's disease, evaluation of stable focal liver disease, metabolic syndrome estimation, study and evaluation of Acquired Immunodeficiency Syndrome (AIDS), Parkinsonian tremor, urological oncology, pigmented skin lesion diagnosis, identification of lung nodules, prediction of the result of epilepsy surgery, and assistance in the treatment of heart disease [43].

9.5.2 Convolutional Neural Networks (CNN)

This neural networks group developed and is very effective in fields like image recognition and classification. The neural network has one or more convolutionary layers and is used primarily for image processing, segmentation, and classification, but also for great advances in the area of computer vision, as well as being used in the fields of drug design, *de novo* drug molecule identification, gene expression analysis, and protein engineering. The tool box for the molecular modelers is equipped with potentially game-changing approaches, with the growth of deep learning hypotheses like CNN [44]. CNN applications in drug design are shown in Figure 9.4.

ANN is a type of CNN, which contains learnable weights and bias neurons. The completely linked network is feasible and used in in-house databases with a small number of descriptors, but the ANN can easily lead to overfitting in massive datasets with various network parameters (weights) and descriptors (ChEMBL database). Therefore, to minimize the problem of redundancy, CNN was introduced. Unlike a feed-forward neural network (NN), which is used only in completely connected layers, there are three different layers of CNN architecture: the pooling layer, the convolutionary layer, and the completely connected layer. CNN has K filters (or kernels), which are smaller in size than the image dimension, the convolutionary layers used for the extraction of features from input neurons. On the other hand, CNN will

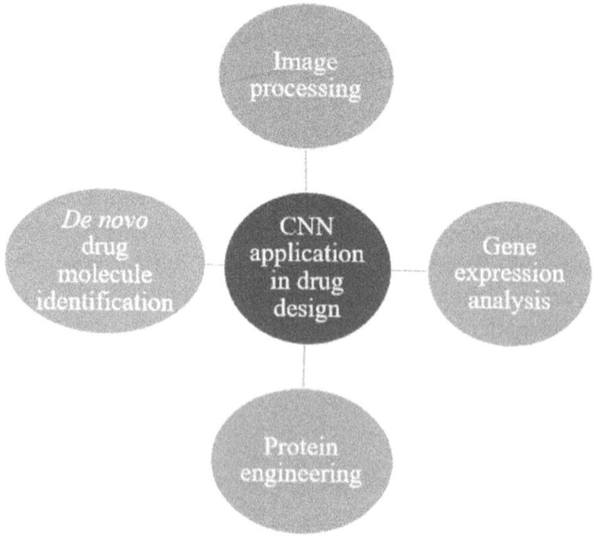

FIGURE 9.4 Schematic diagram of CNN applications in drug design.

measure the output of the neuron linked with local input regions. The completely linked layer is identical to feed-forward NN layers.

By winning the ImageNet Large Scale Visual Recognition Challenge (ILSVRC), a deep CNN also enhanced deep learning, which became a CNN focal point.. CNN has dominated this challenge in terms of the object recognition field and general image detection. The basic framework for a vast range of applications includes image comprehension, speech recognition, analysis of video, and Artificial Intelligence Comprehension language. Within a convolutionary network, the two guiding principles, namely local connectivity and weight sharing, allow features for the topologically organized input data detection. Additional use of pooling and the involvement of several layers helps the network to study more about the features of the abstract with every additional feature. The input data match the image, the filter matches the weight connections, and the function map (activation map) matches the first hidden layer. In comparison to a completely attached one, hidden-layer neurons are only attached to a local patch within the input image.

CNN is a mathematical model usually consisting of three types of layers: pooling, convolutionary and layers which are fully connected. The first two, the pooling and convolutionary layers, investigate the features of extraction, whereas the third is a completely connected layer, extracting map characteristics, and is shown in the final output. Pixel values are stored in a 2D matrix in the form of digital images, such as a numbers array and a small parameter matrix called the kernel, and an optimizable extractor is used at each image location, making CNN highly effective for processing of an image.

A convolutionary layer intends to find the distinctive shapes, points, and other features, which is visual in nature. It provides insights into the parameters of specific filter operators called convolutions. Kernels mimic the visual properties extracted, like colours and edges, close to the visual cortex. This method is carried out with the use of filter banks. The values of the image on this moving grid are added by using the filter weights. Multiple filters are applied by the convolutionary layer and several maps of functions are produced. Convolutions are a CNN central component and are key to the success of tasks like segmentation and image processing.

For instance, in a picture, the first layer may be used to recognize corners and lines, the next one may assemble these characteristics into various shapes, until the final layer recognizes something like a high-level complex, such as the breed of dog. As epitomized by the GoogLeNetwinning entry into the ImageNet Large Scale Visual Recognition Challenge and the entry of Microsoft ResNet, CNN is the best-performing image recognition method. The protein ligand scoring model of CNN is used to identify the various compounds. The 3D grid is used for the representation of structures of the protein-binding ligand that are formed by docking, with the help of binders or nonbinders. It shows that the AutoDockVina scoring algorithm is used to achieve a prediction, which is performed by the CNN scoring process [45].

9.6 APPLICATION OF COMPUTER-AIDED TECHNIQUES TO PHARMACEUTICAL EMULSION DEVELOPMENT

Spreading systems consist of two nonscatterable fluids, forming emulsions. In the presence of surface-action agents like emulsifiers, one substance changes into the other form. The two immiscible fluids, such as water and oil, are used in the emulsion preparation, while water in oil (w/o) or oil in water (o/w) are the main types of clear emulsion. Emulsions have great potential in pharmaceuticals, such as gravity (cream/sediment) separation, flocculation, coalescence, Ostwald maturation, and the reversal of the process. More intensive use of different silicon technologies began at the end of the 20th century in the production process and formulation optimization. In 1998, Prinderre *et al* [46]. used methods of factory design to increase the o/w emulsion stability, which is prepared with a surfactant such as sodium lauryl sulfate, and proposed the hydrophilic-lipopyrene balance (HLB). The independent variables and their rates (lower/higher) were revolutions per minute (500/900) and mixing time (10/20 min). Their variables were homogenized. The average droplet size, emulsion viscosity, and emulsion conductivity are the dependent variables. Experimental design calculated the HLB needed for the average viscosity and diameter trials, with a good approximation from five runs, whereas the conductivity analysis took at least eight runs.

Due to complicated nonlinear correlation problems, simultaneous analysis of the effects of different factor emulsion systems is difficult. A useful method to solve these problems is the methodology of ANN. Using this method, Gasperlin *et al.* [25]determined the effect of various proportions of the individual components on the viscoelastic activity of semi-solid lipophilic emulsion systems. The creams were designed experimentally (mixture design). Three inputs, 12 secret neurons, and nine output neurons were included in ANN. The input neurons were calculated at various time intervals by complex rheological parameter values and by the contents of specific emulsion compounds (silicone surfactant Abil® WE 09, purged water, white petrolatum) and output neurons. A neural network model might be expected in Tanδ, and the results were shown to be highly reliable. These semi-solid w/o emulsions may likewise be estimated by a neural network pattern to achieve complex dynamic viscosity.

The most popular emulsions are w/o/w, although o/w/o emulsions can also be made for specific applications. The two-stage production process usually produces multiple emulsions, that include one method for the manufacture of the primary emulsion and one for secondary emulsions. Oil globules with small water droplets are spread in a continuous aqueous phase in w/o/w emulsions. Such types of emulsion systems have a relative advantage in terms of hydrophilic compound trapping capability, the safety of encapsulated substances to dissolve, the ability to introduce incompatible substances into the same system, and a continuous active substance release. These characteristics are used to add a variety of factors to w/o/w emulsion applications, including external and internal water phase osmotic balance, phase volume ratio, preparation process, and emulsifier concentration.

Onuki *et al.* [47] used statistical methods, like surface assessment and response to the surface, to maximize w/o/w multiple emulsions with insulin. According to the orthogonal experimental design, 16 types of emulsion formulations were prepared. The influences of five variables on emulsion properties were evaluated first to optimize the formulation. The variables were gelatin, insulin, oleic acid, outer-water phase volume ratio, and second emulsion agitation. Exit sizes, viscosity, stability, and pharmacological performance were internal variables. Based on the Analysis of Variance (ANOVA), the outer-water phase volume ratio became the main contributor to all causal factors. A simultaneous optimization technique was employed to estimate optimal formulation for the pharmacological hypoglycaemic action in rats and the emulsion stability by using a multivariate spline interpolation (MSI). The selection of model formulations was performed using a hybrid two-factor secondary spherical experimental design. The authors stated that the optimal formulation had the highest pharmaceutical activity and stability [47].

9.7 APPLICATION OF COMPUTER-AIDED TECHNIQUES TO THE MICROEMULSION DRUG CARRIER DEVELOPMENT

The microemulsion system consists of water, oil, and surfactant, and is thermodynamically stable with optically isotropic, transparent, colloidal structures. Although these liquids are transparent and viscous, different kinds of microstructures (i.e. o/w, w/o, bi-continuous) are known, all of which are organized at a level below 100 nm. The physicochemical properties and concentrations of constituents affect the microemulsion microstructure. Interest is increasing in such special systems, as they show potential for medication delivery with long-term durability, substantial drug solubilization capability, and a great potential to improve bioavailability, in so-called self-microemulsifying oil/surfactant blends.

This carrier architecture involves a complicated strategy that incorporates all important aspects. Several coarse dispersions and colloidal systems may be developed in a system consisting of oil and water (e.g. self-microemulsifying drug delivery systems [SMEDDS]) in a well-balanced excipient mixture at a prescribed temperature and pressure (i.e., the microemulsion field) in a specific range of the constituents. During phase behavioral studies, the effect of formulation variables in the microemulsion field is typically analyzed. Microemulsions available from pharmaceuticals are composed of five (surfactant, co-surfactant, oil, water, and drug) or more components. In these multicomponent mixtures, a huge number of experiments are used to complete the stage distinction behavior study.

However, because of its dynamic nature and the organization of a nanoscale, microstructure characterization is a difficult task. The pioneer studies by Robert *et al.* [39] demonstrated that ANN is used to classify the physicochemical characteristics in a system of four components, in which isopropyl myristate is used as an oil, lecithin as the surfactant, triple distilled water, and the cosurfactant for the formation of the microemulsion. Various types of cosurfactant compounds were used (e.g. medium- and short-chain alcohols, ethylene glycol monoalkyl ethers, and acids). Computed surface molecular properties measure the value of octanol and water logP including molecular volume (v), head group areas ($a\psi$) and hydrophobic elements ($a\varphi$). ANN models have been implemented as useful instruments for distinguishing and estimating the quantitative and qualitative composition of the various microemulsion-forming systems of the microemulsion field [48].

9.8 APPLICATIONS OF MULTISCALE METHODS IN DRUG DISCOVERY

Pharmacology is concerned with a broad range of time and size scales, particularly in terms of drug design, drug development, and drug delivery. Simulation methods, multiscale computational modeling and paradigms advance the connection of phenomena occurring at various scales.

9.8.1 Approaches of Multiscale Modeling

- Efficientsystem and application in cardiac disease.
- Complex biosystems driven by stem cells.
- Delivery of nanoparticles with applications in cancer and angiogenesis therapy.
- Host-pathogen interaction and use in treating inflammation and disorders of metabolism.
- Design of nanomedical systems with CADD [49].

9.8.1.1 Cardiac Modeling Molecular Dynamics

Multiscale drug modeling is an interesting method, since studies on a scale of a single system cannot recognize the interactions associated with multiple drug effects. It is used in disciplines like diseases screening, research and development, academia, and medical clinics. The far-reaching effects are related to the computational-based approach because the increased risk of stratification of drug-based treatments will help millions of people suffering from cardiac arrhythmia every year. The development of multiscale computational modeling and simulation methods to predict the effects of a drug that blocks the cardiac ion channel has made a lot of progress. A critical approach for current and future attempts to discover drugs is the structural modeling for drug interactions between the ion channels. Cardiac ion channel–drug interactions were modeled on an atomic scale in MD simulations and simulated molecular docking, as well as with respect to channel function to simulate channel effect on the drug [50].

9.8.1.2 Network Biology and Cancer Modeling

Cancer is the most dangerous disease today, the World Health Organization has declared. The complexities of cancer progression may be manifested on at least three scales, which may be investigated by using mathematical approaches, such as microscopic, macroscopic, and mesoscopic scales. Multiscale cancer approaches are useful because they can integrate various aspects at the same time. This can guide the development of more data-driven models from predictive models, combined with clinical and experimental data. The foundations that promote the potential design of effective cancer therapies for patients are represented by these models. This can be interpreted as a major move into the realm of personalized medicine. An additional advance in mathematical modeling will lead to more advanced approaches to cancer therapy [51].

9.8.1.2.1 Approaches toward a Dataset Model in Cancer

9.8.1.2.1.1 Perturbation Theory Machine Learning-Linear Discriminate Analysis (PTML-LDA Model) Models based on the ChEMBL dataset are used to determine preclinical compounds for anticancer Activity. The PTML model combines the perturbation theory (PT) concepts with machine learning (ML) techniques to address related problems [52].

9.8.1.2.1.2 Multi-Task (mtk) Chemical Information Model From the dataset of 1,933 peptides, the mtk model incorporates the Moreau-Broto autocorrelation with ANN. This model was used to screen peptides for anti-cancer activity, with virtual design against various cancer cell lines. Wang *et al.* [53] carried out various experiments that show how multiscale models are used for drug target combination and identification therapy. This model is based on the relationship between intracellular epidermal growth factor receptor (EGFR) signaling kinetics, stimulation of extracellular epidermal growth factor (EGF) in lung cancer, and multicellular growth quantification [53].

9.8.1.2.1.3 Multiscale Modeling (MSM) for Discovery of Drugs Used in Brain Disease Treatment Some models focus on the level of the neural network, including the models developed by Moustafa and Cutsuridis [54] on Alzheimer's disease, and by Lytton [55] on epilepsy. The approaches to MSM neuronal network modeling are significantly different from those associated with ANN. Machine learning tools are also used in drug development. Machine learning approaches are utilized in clinical big data analysis which includes the big data associated with drug trials. Computer simulations are used to study brain disorders and diseases. Models may be used directly to estimate the effects of particular drugs planned to be used in animal models to validate their efficacy before being used in the animal study itself. For precision medicine, models can be used because they target known subpopulations that have a specific disease subtype, typically based on their genetics. In diseases like Alzheimer's disease, models can also recognize biomarkers. Proper diagnosis is essential for proper care. Models can be used for the identification and treatment of subtypes of a specific disease. Certain brain disorders do not represent a uniform entity but exhibit different pathologies of a common pathway [56].

9.8.1.2.1.4 Transmissible Diseases Hepatitis C, AIDS, influenza, and other viral diseases, such as the novel coronavirus COVID-19, are transmissible diseases. The viruses' unique structure and mode of proliferation pose a natural challenge to the discovery of effective drugs. Viruses do not possess a cellular structure and autonomous metabolism, but they must proliferate and replicate in the host cells. Some antiviral drugs serve as a viral replication inhibitor. Therefore, drug research includes the implementation of a multiscale model for testing drugs which prevent viral replication, while minimizing damage to the body of the patient.

9.8.1.2.1.5 Retroviral Infections The Human Immunodeficiency Virus (HIV) causes an incurable disease. ChEMBL are used to determine the capabilities of big data obtained from complex datasets that difficult to interrogate because these datasets explain the various characteristics to predict new drug activity against retroviral infections. A PTML model for the ChEMBL dataset has been used effectively for antiretroviral compounds in preclinical experimental analysis. The PT operator is based on a multiconditional system that associates with various simplifications and functions of data management. Due to its various features, combined with preclinical experimental antiretroviral tests, the PTML model was proposed and considered first to study Gram-negative bacteria and *in-vitro* safety associated with antibacterial activity against ADMET. The best model demonstrated accuracies in both prediction (testing sets) and training sets of around 93%. Additionally, by using this model, the accuracy of the anti-hepatitis C virus (HCV) in the prediction and training sets was greater than 95%. Cytotoxicity is the main problem in the primary production of drugs based on a peptide. Kleandrova *et al.* [57] investigated the first computational multitask processing (mtk) model based on the prediction of both peptide cytotoxicity and antibacterial activity. Gonzalez-Diaz *et al.* estimated the Linear Neural Network-Assessing Link with Moving Averages (LNN-ALMA) model to produce complex networks related to AIDS occurrence, which was linked to the preclinical behavior of anti-HIV drugs. Multiscale models appear to be incomplete and have disadvantages. Models will assist in real-life expressions and simplifications. No model can represent everything in the system that is happening. All models include particular expectations and differ greatly in their consistency, utility, and comprehensiveness. Only limited issues can be solved by a single model [57].

Pharmaceutical efficiency has faced obstacles amid tremendous success in biomedical science and technology, as well as increased research and development expenses. Success rates in the development and production of new drugs remain poor and, in the past decades, has shown a decreasing trend. The inefficiency of drug production has also been recognised by the US Food and Drug Administration (FDA), who suggested model-based drug development (MBDD) to enhance pharmaceutical effectiveness and decision making. Modeling and simulation offer a good method for summarizing and combining data from various studies. The use of simulation can assist with decision making, can help to plan better experiments, minimise costs, save time, and eventually increase success rates. MBDD is a paradigm that encompasses the whole range of the drug development process, beyond conventional types of modeling techniques or applications.

9.9 ACCELERATED DRUG DEVELOPMENT BY MACHINE LEARNING METHODS

Machine learning (ML) is the process of using algorithms to analyze the data, learn from it, and then evaluate and predict the future conditions of any new datasets. This provides a chance to learn the technique to perform the task. Instead of rules from coding experts, programmer codes are used to train the network. The algorithms adaptatively enhance their efficiency as the quality and quantity of data available for learning increases. Therefore, ML is used to solve problems for a huge volume of data, with many variables available, but the formulae or model related to these is unknown. There are two types of techniques used for ML application, namely supervised and unsupervised learning. Supervised learning techniques are used to elaborate training models to determine the future data category values or continuous variables, while unsupervised methods are employed to develop models for the exploratory purposes that enable data to be clustered in a manner not specified by the consumer. The unsupervised learning method detects hidden patterns or intrinsic structures in the input data and makes practical use of these to cluster data.

With high-quality data from well-specified problems, ML methods offer a variety of resources which can enhance decision- and discovery-making properties. Examples include validation of the target, prognostic biomarker identification, and digital pathology data analysis in an animal study. Applications range from methodology and context, with some methods generating precise estimations and insights. The difficulties of applying ML are mainly due to the absence of interpretability and repeatability of results provided by ML, which can restrict their application.

In the pharmaceutical industry, ML approaches are most widely used for virtual compound screening. ML algorithms are widely used in both toxicological and pharmaceutical research. In electronic health records, ML approaches are used to accurately predict multiple medical events from various centers without specific site data harmonization, with recent data suggesting that profound learning was comparable to regular logistic regression [58].

9.10 CONCLUSION

CADD plays a necessary role in drug development and drug delivery. It helps in the development of drugs in less time with lower costs and labor. It helps in the optimization of the formulation, which makes it easier to design the perfect formulation of the drug without any problem. Different approaches, such as molecular dynamics and molecular docking, are the major keys to the CADD process. Deep learning methods, like ANN and CNN, are also used in drug development. CADD can also facilitate the medical field with the diagnoses of various diseases, making it easier to diagnose the disease accurately in less time. In the future, CADD will increase research quality and will be used in the discovery of novel formulations of drugs.

REFERENCES

1. Gupta, R., Rai, B., & Kewal, K. J. (Eds.).(2020). Computer-aided design of nanoparticles for transdermal drug delivery. In *Drug Delivery Systems* (pp. 225–237). Humana, New York, NY.
2. Ramezanpour, M., Leung, S.S.W., Delgado-Magnero, K.H., Bashe, B.Y.M., Thewalt, J., & Tieleman, D.P. (2016). Computational and experimental approaches for investigating nanoparticle-based drug delivery systems. *Biochimica et Biophysica Acta (BBA) - Biomembranes, 1858*(7), 1688–1709.
3. Nandy, A., & Basak, S.C. (2016). A brief review of computer-assisted approaches to rational design of peptide vaccines. *International Journal of Molecular Sciences, 17*(5), 1–11.
4. Ebhohimen, I.E., Edemhanria, L., Awojide, S., Onyijen, O.H., & Anywar, G., & Egbuna, C., Kumar, S., Ifemeje, J. C., Ezzat, S. M., Kaliyaperumal, S. (Eds.).(2020). Advances in computer-aided drug discovery. In *Phytochemicals as Lead Compounds for New Drug Discovery* (pp. 25–37). Elsevier.
5. Yu, W., & MacKerell, A.D., & Sass, P. (Eds.). (2017). Computer-aided drug design methods. In *Antibiotics* (pp. 85–106). Humana Press, New York, NY.
6. Tomar, V., Mazumder, M., Chandra, R., Yang, J., & Sakharkar, M. K., & Trabocchi, A., Lenci, E. (Eds.). (2019). *Small Molecule Drug Design.* In *Reference Module in Life Sciences* (pp. 741–760). Elsevier.
7. Li, Q., Shah, S., Wu, H.C., Arighi, C.N., & Ross, K.E. (Eds.). (2017). Structure-based virtual screening. In *Protein Bioinformatics* (pp. 111–124). Humana Press, New York, NY.
8. van Montfort, R.L., Workman, P., Lamoree, B., & Hubbard, R.E. (2017). Current perspectives in fragment-based lead discovery (FBLD). *Essays in Biochemistry, 61*(5), 453–464.
9. Ferreira, L.L., & Andricopulo, A.D. (2019). ADMET modeling approaches in drug discovery. *Drug Discovery Today, 24*(5), 1157–1165.
10. Macalino, S.J., Gosu, V., Hong, S., & Choi, S. (2015). Role of computer-aided drug design in modern drug discovery, *Archives of Pharmacal Research, 38*, 1686–1701.
11. de Queiroz Simoes, R.S., Ferreira, M.S., de Paula, N.D., Machado, T.R., Pascutti, P.G., da Silva F.A.B., Carels N., Trindade dos Santos M., & Lopes F.J.P. (Eds.) (2020). Computational modeling in virus infections and virtual screening, docking, and molecular dynamics in drug design. In *Networks in Systems Biology* (pp. 301–337). Springer, Cham.
12. Yu, W., MacKerell, A.D., & Sass, P. (Eds.). (2017). Computer-aided drug design methods. In *Antibiotics* (pp. 85–106). Humana Press, New York, NY.

13. Hung, C.L., & Chen, C.C. (2014). Computational approaches for drug discovery. *Drug Development Research, 75*(6), 412–418.
14. Macalino, S.J., Gosu, V., Hong, S., & Choi, S. (2015). Role of computer-aided drug design in modern drug discovery. *Archives of Pharmacal Research, 38*(9), 1686–1701.
15. Sliwoski, G., Kothiwale, S., Meiler, J., & Lowe, E.W. (2014). Computational methods in drug discovery. *Pharmacological Reviews, 66*(1), 334–395.
16. Yang, S.Y. (2010). Pharmacophore modeling and applications in drug discovery: Challenges and recent advances. *Drug Discovery Today, 15*(11–12), 444–450.
17. Alayoubi, A., Satyanarayanajois, S.D., Sylvester, P.W., & Nazzal S. (2012). Molecular modelling and multisimplex optimization of tocotrienol-rich self-emulsified drug delivery systems. *International Journal of Pharmaceutics, 426*, 153–161.
18. Tao, X., Huang, Y., Wang, C., Chen, F., Yang, L., Ling, L., Che, Z., & Chen, X. (2020). Recent developments in molecular docking technology applied in food science: A review. *International Journal of Food Science & Technology, 55*(1), 33–45.
19. Zhao, H., & Caflisch, A. (2015). Molecular dynamics in drug design. *European Journal of Medicinal Chemistry, 91*, 4–14.
20. Kerrigan, J.E., & Kortagere, S. (Eds.). (2013). Molecular dynamics simulations in drug design. In *In Silico Models for Drug Discovery* (pp. 95–113). Humana Press, Totowa, NJ.
21. Gaba, M., Gaba, P., Singh, S., & Gupta, G.D. (2010). An overview on molecular docking. *International Journal of Drug Development & Research, 2*, 219–231.
22. Eweas, A.F., Maghrabi, I.A., & Namarneh, A.I. (2014). Advances in molecular modeling and docking as a tool for modern drug discovery. *Der Pharma Chemica, 6*, 211–228.
23. Kumar, S., Purohit, D., Pandey, P., & Neeta (2017). Molecular Docking and its application towards Modern Drug Discovery. *World Journal of Pharmacy and Pharmaceutical Sciences, 6*, 691–696.
24. Suresh, P.S., Kumar, A., Kumar, R., & Singh, V.P. (2008). An *in silico* [correction of *insilico*] approach to bioremediation: Laccase as a case study. *Journal of Molecular Graphics and Modelling, 26*, 845–849.
25. Gasperlin, M., Podlogar, F., & Sibanc, R. (2008). Evolutionary artificial neural networks as tools for predicting the internal structure of microemulsions. *Journal of Pharmacy and Pharmaceutical Sciences, 11*, 67–76.
26. Hussain, A.S., Johnson, R.D., Vachhrajani, N., & Ritschel, W.A. (1993). Feasibility of developing a neural network for prediction of human pharmacokinetic parameters from animal data. *Pharmaceutical Research, 10*, 466–469.
27. Aghajani, M., Shahverdi, A.R., & Amani, A. (2012). The use of artificial neural networks for optimizing polydispersity index (PDI) in nanoprecipitation process of acetaminophen in microfluidic devices. *AAPS PharmSciTech, 13*(4), 1293–1301.
28. Ebube, N.K., Owusu-Ababio, G., & Adeyeye, C.M. (2000). Preformulation studies and characterization of the physicochemical properties of amorphous polymers using artificial neural networks. *International Journal of Pharmaceutics, 196*(1), 27–35.
29. Onuki, Y., Kawai, S., Arai, H., Maeda, J., Takagaki, K., & Takayama, K. (2012). Contribution of the physicochemical properties of active pharmaceutical ingredients to tablet properties identified by ensemble artificial neural networks and kohonen's self-organizing maps. *Journal of Pharmaceutical Sciences, 101*(7), 2372–2381.
30. Sutariya, V., Groshev, A., Sadana, P., Bhatia, D., & Pathak, Y. (2013). Artificial neural network in drug delivery and pharmaceutical research. *The Open Bioinformatics Journal, 7*(1), 49–62.
31. Agatonovic-Kustrin, S., Zecevic, M., & Zivanovic, L.J. (1999). Use of ANN modelling in structure–retention relationships of diuretics in RP-HPLC. *Journal of Pharmaceutical and Biomedical Analysis, 21*(1), 95–103.

32. Ibric, S., Djuris, J., Parojcic, J., & Djuric, Z. (2012). Artificial neural networks in evaluation and optimization of modified release solid dosage forms. *Pharmaceutics*, *4*(4), 531–550.
33. Peh, K.K., Lim, C.P., San Quek, S., & Khoh, K.H. (2000). Use of artificial neural networks to predict drug dissolution profiles and evaluation of network performance using similarity factor. *Pharmaceutical Research*, *17*(11), 1384–1389.
34. Mendyk, A., Tuszynski, P.K., Polak, S., & Jachowicz, R. (2013). Generalized in vitro-in vivo relationship (IVIVR) model based on artificial neural networks. *Drug Design, Development and Therapy*, *7*, 223–232.
35. Mandlik, V., Bejugam, P.R., Singh, S., Tipparaju, S., Moreno, W., Pathak, Y., & Puri, M. (Eds.). (2016). Application of artificial neural networks in modern drug discovery. In *Artificial Neural Network for Drug Design, Delivery and Disposition* (pp. 123–139). Academic Press.
36. Bala, R., & Kumar, D. (2017). Classification using ANN: A review. *International Journal of Computational Intelligence Research*, *13*(7), 1811–1820.
37. Rogers, J.C., Qu, Y., Tanada, T.N., Scheuer, T., & Catterall, W.A. (1996). Molecular determinants of high affinity binding of α-scorpion toxin and sea anemone toxin in the S3-S4 extracellular loop in domain IV of the Na+ channel α subunit. *Journal of Biological Chemistry*, *271*(27), 15950–15962.
38. Chen, B., Xia, Z., Deng, Y.N., Yang, Y., Zhang, P., Zhu, H., Xu, N., & Liang, S., (2019). Emerging microRNA biomarkers for colorectal cancer diagnosis and prognosis. *Royal Society Open Biology*, *9*(1), 1–11.
39. Lancashire, L.J., Lemetre, C., & Ball, G.R. (2009). An introduction to artificial neural networks in bioinformatics—Application to complex microarray and mass spectrometry datasets in cancer studies. *Briefings in Bioinformatics*, *10*(3), 315–329.
40. Lin, B., Lin, G., Liu, X., Ma, J., Wang, X., Lin, F., & Hu, L. (2015). Application of back-propagation artificial neural network and curve estimation in pharmacokinetics of losartan in rabbit. *International Journal of Clinical and Experimental Medicine*, *8*(12), 22352–22358.
41. Garg, P., & Verma, J. (2006). In silico prediction of blood brain barrier permeability: An artificial neural network model. *Journal of Chemical Information and Modeling*, *46*(1), 289–297.
42. Chen, L.J., Lian, G.P., & Han, L.J. (2007). Prediction of human skin permeability using artificial neural network (ANN) modeling 1. *Acta Pharmacologica Sinica*, *28*(4), 591–600.
43. Al-Shayea, Q.K. (2011). Artificial neural networks in medical diagnosis. *International Journal of Computer Science Issues*, *8*(2), 150–154.
44. Krizhevsky, A., Sutskever, I. & Hinton, G.E., Hanson, S.J., Cowan, J.D., & Giles, C.L. (Eds.). (2012). Imagenet classification with deep convolutional neural networks. In *Advances in Neural Information Processing Systems* (pp. 1097–1105), Morgan Kaufmann Publishers Inc., 340 Pine Street, Sixth Floor, San Francisco, CA, United States.
45. Ragoza, M., Hochuli, J., Idrobo, E., Sunseri, J., & Koes, D. R. (2017). Protein–ligand scoring with convolutional neural networks. *Journal of Chemical Information and Modeling*, *57*(4), 942–957.
46. Prinderre, P., Piccerelle, P., Cauture, E., Kalanis, G., Reynier, J.P., & Joachim, J. (1998). Formulation and evaluation of o/w emulsions using experimental design. *International Journal of Pharmaceutics*, *163*, 73–79.
47. Onuki, Y., Morishita, M., & Takayama, K. (2004). Formulation optimization of water-in-oil-water multiple emulsion for intestinal insulin delivery. *Journal of Controlled Release*, *97*, 91–99.

48. Djekic, L., Ibric, S., & Primorac, M. (2008). The application of artificial neural networks in the prediction of microemulsion phase boundaries in PEG-8 caprylic/capric glycerides-based systems. *International Journal of Pharmaceutics*, *361*, 41–46.
49. Clancy, C.E., An, G., Cannon, W.R., Liu, Y., May, E.E., Ortoleva, P., Popel, A.S., Sluka, J.P., Su, J., Vicini, P., & Zhou, X. (2016). Multiscale modeling in the clinic: Drug design and development. *Annals of Biomedical Engineering*, *44*(9), 2591–2610.
50. Moreno, J.D., Zhu, Z.I., Yang, P.C., Bankston, J.R., Jeng, M.T., Kang, C., Wang, L., Bayer, J.D., Christini, D.J., & Trayanova, N.A. (2011). A computational model to predict the effects of class I anti-arrhythmic drugs on ventricular rhythms. *Science Translational Medicine*, *3*, 83–98.
51. Masoudi-Nejad, A., & Wang, E. (Eds.). (2015). Cancer modeling and network biology: Accelerating toward personalized medicine. In *Seminars in Cancer Biology* (pp. 1–3). Academic Press.
52. Bediaga, H., Arrasate, S., Gonzalez-Diaz, H. (2018). PTML combinatorial model of ChEMBL compounds assays for multiple types of cancer. *ACS Combinatorial Science*, *20*, 621–632.
53. Kim, M., Gillies, R.J., & Rejniak, K.A. (2013). Current advances in mathematical modeling of anti-cancer drug penetration into tumor tissues. *Frontiers in Oncology*, *3*, 1–10.
54. Cutsuridis, V., & Moustafa, A. (2016). Multiscale models of pharmacological, immunological and neurostimulation treatments in Alzheimer's disease. *Drug Discovery Today: Disease Models*, *19*, 85–91.
55. Lytton, W. (2016). Computer modeling of epilepsy: Opportunities for drug discovery. *Drug Discovery Today: Disease Models*, *19*, 27–30.
56. Neymotin, S.A., & Lytton, W.W. (2016). Multiscale modeling for drug discovery in brain. *Drug Discovery Today: Disease Models*, *19*, 1–3.
57. Kleandrova, V.V., Ruso, J.M., Speck-Planche, A., Dias SoeiroCordeiro, M.N. (2016). Enabling the discovery and virtual screening of potent and safe antimicrobial peptides. simultaneous prediction of antibacterial activity and cytotoxicity. *ACS Combinatorial Science*, *18*, 490–498.
58. Johnson, K.W., Shameer, K., Glicksberg, B.S., Readhead, B., Sengupta, P.P., Bjorkegren, J.L., Kovacic, J.C., & Dudley, J.T. (2017). Enabling precision cardiology through multiscale biology and systems medicine. *JACC: Basic to Translational Science*, *2*(3), 311–327.

10 Healthcare Data Analytics Using Business Intelligence Tool

Annapurani K., E. Poovammal, C. Ruvinga, and Ibrahim Venkat

CONTENTS

10.1 INTRODUCTION: BIG DATA

Big data is a huge collection of data from different sources. The data collected can be structured, unstructured, or semi-structured. The processing speed has to be very high and the results are expected immediately. Data play a crucial role in obtaining the insights from the information for implementing strategic business decisions. Usually, enormous volumes of data are needed to obtain useful business insights. This huge amount of data is termed "big data", in the range of petabytes or exabytes, that cannot be handled by traditional methods or technologies.

Big data is characterized by Variety, Volume, Velocity, Veracity, and Value, as shown in Figure 10.1. *Variety* denotes the type and nature of the data. Big data consists of heterogeneous sources of data, with varying degrees of complexity, such as text, audio, video, social media, Internet of Things (IoT), etc. *Volume* is the size or quantity of the data stored, which ranges from yottabyte, zettabyte, and exabyte to petabyte. The storage of huge quantities of data is enabled by advanced technology and scalable infrastructure. *Velocity* is the continuous rate at which data is generated and processed to meet client demand or organizational need. This can be achieved through improved connectivity technology, e.g., batch, real-time, or stream processing. *Veracity* refers to the quality of the data, its origin, accuracy, completeness, and clarity. The data must be traceable and justifiable for the purpose of accurate analysis. *Value* is the consequence of transforming big data into useful information for strategic business decision making and accurate predictions.

Data collected from heterogeneous sources can be structured, unstructured, or semi-structured [1]. Structured data have a distinct storage format, for example, with rows and columns in a relational database, or a table with client details, such as id, name, address, and contact number. Unstructured data have no defined representation, e.g., pdf, text documents, images, videos, etc.

Semi-structured data have a visible pattern that facilitates analysis, such as XML files and spreadsheets. The heterogeneous data must be processed with the aim of

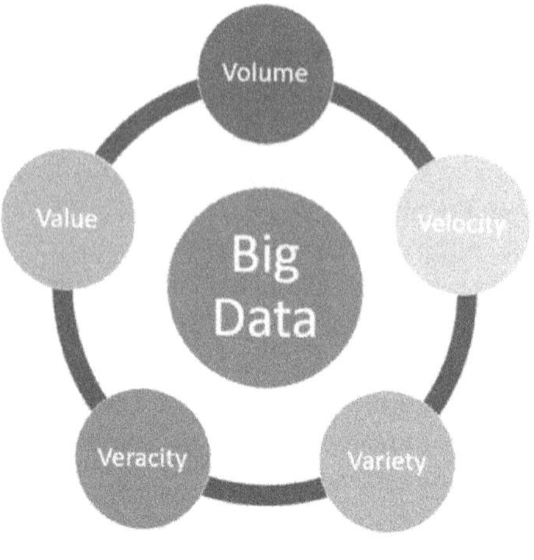

FIGURE 10.1 Big data characteristics (adapted from vapulus.com).

obtaining valuable information. The modern era requires speedy data processing into useful information for implementation of decisions at individual, group, and organizational level. The knowledge obtained from the information helps the decision makers to implement the right decision at the right time and place. This information provides an insight to arrive at accurate strategic decisions for the managers or organizations to succeed. The enormous volume of data from different sources that are processed rapidly enough describes the characteristics of information [2]. This chapter discusses the healthcare-related data and its effects.

Data analytics are of paramount importance to diverse fields, including the healthcare sector. Healthcare data analytics transforms healthcare data into useful information for various components within and related to the healthcare sector. The plethora of components ("actors") consists of doctors, nurses, and healthcare facilities, such as hospitals, diagnostic laboratories, and financial institutions, such as banks, that provide the necessary support for the establishment of healthcare facilities. The useful information acquired from the healthcare data analytics system is derived from enormous volumes of healthcare data. The healthcare data used by the healthcare analytics systems are drawn from heterogeneous sources, such as patient laboratory test reports, consultation records, e-prescriptions, pharmacy billing reports, review reports, medical diagnosis reports, and healthcare research.

The useful insights aid healthcare providers in effectively treating future patients suffering from similar or related medical conditions, according to their need and the level of emergency [3]. For example, healthcare data analytics can assist in identifying or categorizing patients at risk of certain chronic illnesses, like cancer and diabetes. These insights enable implementation of accurate time decisions by healthcare components. Healthcare data analytics are also useful in obtaining aggregate patient data from multiple sources, which, in turn, provide for more timely, informed patient treatment, and increased satisfaction and comfort achieved by healthcare providers. In a nutshell, healthcare data analytics can be viewed as being important for the overall maintenance and improvement of healthcare service delivery.

In this respect, data collection is an important phase in any data analytics research. Currently, there is an outbreak of the disease COVID-19, which is still new and against which the proper countermeasures are still to be found. However, we can make use of the datasets that are available to predict the extent of the outbreak, the number of infected people according to gender, and the most highly infected regions. Healthcare data analytics involves the collection of insights from patient datasets, which is usually collected from various sources, such as Electronic Health Records (EHR), prescriptions from doctors, first form filled by the patients themselves, and laboratory reports. These details might not be in the required format for analysis. Hence, they must be processed according to our requirements. For example, the name, age, and the symptoms of disease might be collected in one place, whereas, in another hospital, the name, date of birth, and the symptoms might be stored. When the data are being processed, uniformity must be achieved, with the date of birth being converted to age. With respect to country or domain, the datasets available in data.gov.in are used.

The necessary data analysis is performed and a dashboard created that presents a suitable visualization to help to achieve a better understanding of the datasets and provide insights based on those visualizations. Most of the states affected by the COVID-19 disease have been identified. It would be helpful to isolate the boundaries of these states from the other unaffected or less-affected states. Also, the statistics of active and recovered cases are obtained. These data would be helpful for the medical facility in the states where the recovery rate is poor. We can provide warning to the people to avoid such areas and also to increase the number of healthcare units in that area.

Machine learning is the technique or algorithm used to describe the behavior of the present system, to predict and optimize the system. In the healthcare system, machine learning is used for predicting the infected areas, or the site of outbreak of a pandemic. The tool was not properly applied to the existing data about 100 years ago when the Spanish 'flu pandemic occurred. That pandemic resulted in a massive loss of human life, especially of frontline workers, as well as economic loss across almost all countries in the world. Even the developed nations suffered due to that pandemic since the data were not properly analyzed. This shows the importance of analysis of the existing data to arrive at a clear-cut solution. Insight into any available information is of foremost importance when making a decision. Microsoft Power-BI tools help the management of business intelligence with a solution based on the insight from results analysis. Machine learning adds to the intelligence of the system that could play a major role in the future to improve the performance of the system.

10.2 DATA COLLECTION

Data form the basis on which the big data analytics system is developed. If the data are not authentic or correct, then they will generate incorrect insights, which will lead managers or decision makers into making poor, inaccurate decisions, causing mismanagement, and may result in the collapse of the entire organizational system. Data have to be sufficiently accurate to be processed into useful information. Information is wealth that helps people at all levels to take the necessary decisions and may also be used for predicting the future. The healthcare data can be collected from various authentic sources, which include manual health records, EHR, patient laboratory reports, electroencephalograms (EEG) or electrocardiogram (ECG) device recordings [4], to mention but a few.

10.2.1 Electronic Health Records (EHR)

EHR represent data about the patient that are stored in the system digitally instead of in a paper format. It enables the records to be easily accessible by patients and doctors [5], as well as by researchers (with permission from the concerned parties). Since the data are in digital form, the integration of the EHR with other electronic reports, like laboratory reports, becomes an easy process. The automated integration [6] helps in arriving at decisions swiftly. EHR data have to be maintained with the utmost care to avoid misuse, unauthorized access, or manipulation. If proper security protection is not present, for example, using patient data without permission may expose the patient to an embarrassing or life-threatening situation. EHR is used for various purposes, such as billing information, medicine dispensing [5], and claiming of insurance.

EHR also enables easy access to healthcare data for remote diagnosis of patients. A patient monitoring system [7] can be carried out in any rural area with the support of EHR. The success of the patient health diagnosis relies on the correct EHR. Therefore, care has to be taken during the collection, storage, and protection of the patient's past and present health data and information.

10.2.2 Laboratory and Diagnostic Reports

The bodily parameters of an affected patient, such as body temperature, blood pressure, height, weight, gender, and age, constitute the basic laboratory or diagnostic reports data. But, if the patient is in a critical condition, additional parameters are included, such as laboratory blood tests, urine tests, and scans. Diagnosis of patients and their treatment are found to be improved based on additional information from the laboratory tests and scans [8]. Laboratory tests taken, such as blood tests to give the blood platelet counts and hemoglobin concentration, and urine tests and diagnostic results from a Computerized Tomography (CT) scan, Magnetic Resonance Imaging (MRI), or X-ray help to locate the target area of infection or disease [9]. Mostly, these are the parameters that help to identify the reason for a patient's illness and hence to provide the appropriate treatment accordingly.

10.2.3 Prescriptions

Prescriptions can be handwritten or in electronic form. A handwritten prescription is the medication advised by the doctor to a patient when they visit the hospital as an outpatient. The medicines can be bought in the pharmacy based on the doctor's advice. The drawback to this is that the handwriting of the doctor is not always legible to everyone. Errors might occur in giving the drugs when the details of the drugs written in the prescription are not clear. In e-prescriptions, the drawback with handwritten prescriptions is avoided since there is an electronic exchange of the prescription between the doctor and the pharmacy [10].

E-prescriptions can be readily recorded and followed up on. Easy integration of data is also possible when e-prescriptions are used, benefitting the doctors, patients, and pharmacy staff. These data help each actor in their own perspective in different ways. The patient can monitor their health status and take action accordingly. It will be easy for the doctor to see the record of the patient's last visit, and check for any improvement; if there is no improvement, different medicines can be prescribed. From the point of view of the pharmacy, the staff can analyze the medicines sold every month, identifying the amounts of each drug sold, and accordingly maintain the inventory.

10.2.4 Forms Filled By Patients

Once the patient feels sick, either they go to the doctor or hospital in person or they consult a doctor remotely. In either case, the patient's personal details, such as name, age, gender, medical aid provider and number, reason for consultation, symptoms of the disease, etc., are captured by the healthcare provider. This information is normally captured in a standard form at the admission or reception desk before the consultation.

The form contains basic standard information important for unique identification of the patient and, subsequently, for storage. The form also captures the dates of consultation for review and billing purposes. Some medical aid service providers have a limit to the number of monthly consultations they pay for. Hence, such manual forms capture important dates which are necessary for claiming consultation fees from the medical aid service provider or serve as proof as to why a patient needs to pay cash in the event that they have exceeded the maximum number of consultations allowed per specified period, as guided by medical aid policy. This patient form contains necessary information that helps in retrieval of medical files containing treatment history of the patient if they have previously visited the healthcare facility.

The major challenge of handwritten forms is that some important fields may be left blank and are prone to error since they are handwritten. Forms are hard copies and are not in electronic format, and, therefore, are difficult to integrate with electronic laboratory reports. Data from such sources is also prone to security threats, since physical security is mainly the only source of security for such forms.

10.3 DATA PRE-PROCESSING

Data refers to the raw facts recorded, such as given name, surname, patient_ID, weight, height, age, etc., and is broadly categorized into categorical and numerical data, as shown in Figure 10.2. Categorical data comprises of nominal and ordinal data [11]. Ordinal data have a specific order for labeling the data, for example, recording the stage of an infection on a scale range of 1 to 5. Nominal data is unordered, where there is no significant order to be followed for recording; for instance, a patient's choice of spectacle frame color can be green, black, white, brown, etc., for which order is not relevant. Numeric data consists of discrete and continuous data. Discrete data takes specific numeric values, such as the number of previous abortions, number of previous surgeries, or the number of hospital visits, whereas

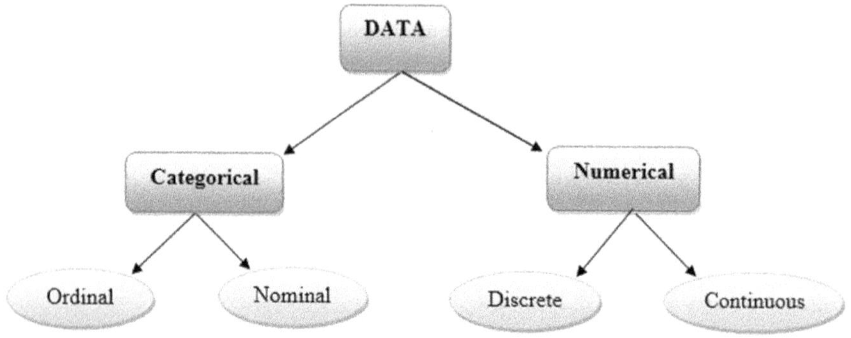

FIGURE 10.2 Data categories (based on [11]).

heartbeat rate or blood pressure are examples of continuous data, i.e., they take non-specific numeric values.

Data pre-processing is performed on "dirty data". Dirty data is characterized by incomplete, noisy, or inconsistent data. Incomplete data lack attribute values and are normally visible in manual and EHR. For example, during data capture, the "number of pregnancies" field may be left blank, intentionally or unintentionally, which results in storage of ambiguous records which cannot be clearly interpreted as having either no pregnancy recorded or as missing data.

Noisy data normally arise as a result of equipment malfunction, such as faulty EEG or ECG readings, resulting in biased or inaccurate recordings of brain waves or heart rate readings, respectively. Blurred medical images from CT scans, MRI scans or X-rays can also be a source of noisy data. Semi-structured data sources, such as medical healthcare spreadsheets, may capture text values instead of numeric values and this gives rise to noisy data, for example, in the "number of hospital visits" column; if NIL is recorded, instead of the digit zero (0), we end up with noisy data. Outliers are another source of noisy data, such as an abnormal overweight outside the scope of the previously recorded medical history range. Deletion or capturing of incorrect patient details may lead to inconsistencies.

Due to these various sources of dirty data, it is imperative to perform data pre-processing before analysis. When pre-processed, the data will lead to precise analysis. If dirty data are used, it will result in GIGO (Garbage In, Garbage Out). There are four main processes for data pre-processing, namely data selection, data cleansing, data transformation, and data integration.

10.3.1 Data Selection

The data selection process retrieves task-relevant data for analysis from heterogeneous data sources, such as structured data sources (e-prescriptions, patient database, laboratory records, pharmaceutical database) and unstructured data sources (CT scans, MRI scans, or X-ray images). Data selection can be implemented using one of numerous data reduction techniques, while maintaining the integrity of the dataset for analysis includes:

- Data cube aggregation combines data. For example, combining the quarterly infant mortality statistics for years 2015–2020 will generate annual infant mortality values for the respective years.
- Attribute/subset selections are normally conducted by a domain expert by removing irrelevant or redundant data. This can be illustrated by capturing patient date of birth (DoB) only, instead of both DoB and age. There is no need to capture age since it can be calculated from DoB.
- Dimensionality reduction involves reducing the number of columns to be used for analysis. The age, weight, and height are relevant for calculation of Body Mass Index (BMI), whereas other parameters, such as the number of previous visits, medical aid policy, name of doctor, etc., are irrelevant and therefore can be excluded from analysis of BMI.

- Sample size reduction refers to using a smaller data representation of the original dataset without compromising the integrity of the original dataset. Instead of using all one million patients' X-ray images, you use a sampling technique to select a specific number of patient records less than the original one million for your analysis. However, the sample should maintain the integrity of the original population.
- Discretization and concept hierarchy generation is the reduction of the number of continuous values recorded by sub-dividing the parameter values into intervals. For instance, instead of using the actual heart rate recordings for all patients, they can be categorized into high, medium, orlow, with the categorical labels being used in place of the continuous values.

10.3.2 Data Cleansing

Data cleansing deals, by removing incomplete, noisy, or inconsistent data [12] from structured, unstructured, or semi-structured data.

- Incomplete (missing) data is typically found in structured data as a lack of attribute values or as having fields with aggregate data only. Data cleansing in such a scenario can be performed by applying an operator that detects missing values, depending on the volume of data. If you have limited data, missing values should be filled in manually, or an operator can be applied which detects missing values and fills in the missing data. A global constant, attribute mean, or the most probable value can be used to replace missing values. Another alternative is to discard the entire record containing the missing data.
- Noisy data are due to various causes, which include random error, outliers, blurred images, incorrect data type entries, etc. Cleansing of noisy data is done by removing or smoothing noisy data, using smoothing techniques, such as Bin equi-depth, Bin mean, or Bin boundaries or clustering. Performing image de-noising also helps to reduce or eliminate salt and pepper, Gaussian, speckle noise, or blurring of medical images. Outliers, known as abnormalities (values outside the expected range), are another source of noisy data. Outliers can be excluded from the analysis as a way of dealing with noise.
- Inconsistent data arise due to non-uniform data type entry within a column. Inconsistency is removed by applying different techniques of making data entries consistent. Some simple techniques used include simple conversion to a similar format, e.g., centimeters to inches in the height attribute, kilograms to pounds for the weight attribute value, and degrees Celsius to Fahrenheit for temperature attribute. In a patient database, the duration attributes for symptoms may be captured in hours in the patient record, whereas, in the laboratory report, it is presented in days. In order to integrate data from the patient database and laboratory report, similar attribute should be expressed on similar attribute value type scales or units. In this scenario, the duration should be presented as either days or hours in both reports.

10.3.3 Data Conversion/Transformation

Data transformation changes data to an appropriate format for analysis. The form in which data are collected may pose challenges to data analytics. For example, working with continuous data may be difficult, and discretization may be necessary. The following techniques are used to perform data transformation: smoothing, aggregation, generalization, normalization, and attribute construction.

- During smoothing, a digital image affected by noise, such as a Computerized Axial Tomography (CAT) scan or an X-ray, is convolved with a relevant smoothing filter, such as the arithmetic mean, harmonic, or adaptive filters, to remove the noise component. Smoothing improves image resolution and quality.
- Aggregation refers to summarization or generalization of data. This results in data conversion. For example, all people within the age range of 12–18 years can be categorized as youths, 19–35 years as young adults, and above 45 as senior for ease of interpretation.
- Normalization can be done using a z-score, decimal scaling, or min–max. This process enables the data values to fall within a certain range that facilitates easy, quick computations, analysis, comparison, and data visualization.
- Attribute construction develops new attributes to facilitate analysis; for example, given the weight and height of an individual, the BMI parameter can be constructed, which can aid in analysis of obesity.

10.3.4 Data Integration

Data integration combines data from heterogeneous sources and may give rise to inconsistencies and redundancy. Heterogeneous healthcare records may have different attribute-naming conventions, depending on the designer or developer of the healthcare system in use. But this different naming convention will be referring to the same attribute value. In order to perform data integration on these heterogeneous data sources successfully, there is a need for assistance from a domain expert to make the data consistent. For example, patient identification number may be stored as P_id in the patient's healthcare record and as Patient_id in their pharmaceutical record. In the event that these two data sources have to be integrated, duplicate fields with slightly different naming conventions have to be removed by dropping one of the two columns. Although an attribute name may differ, the column entry refers to the same data.

10.4 DATA ANALYTICS AND BI

Data Analytics and Business Intelligence (BI) are closely related fields. The aim of BI is to create business operational efficiency, whereas data analytics focuses on more mathematical inferences from huge amounts of data. BI refers to the application of business technologies for retrieving relevant, accurate, and interesting

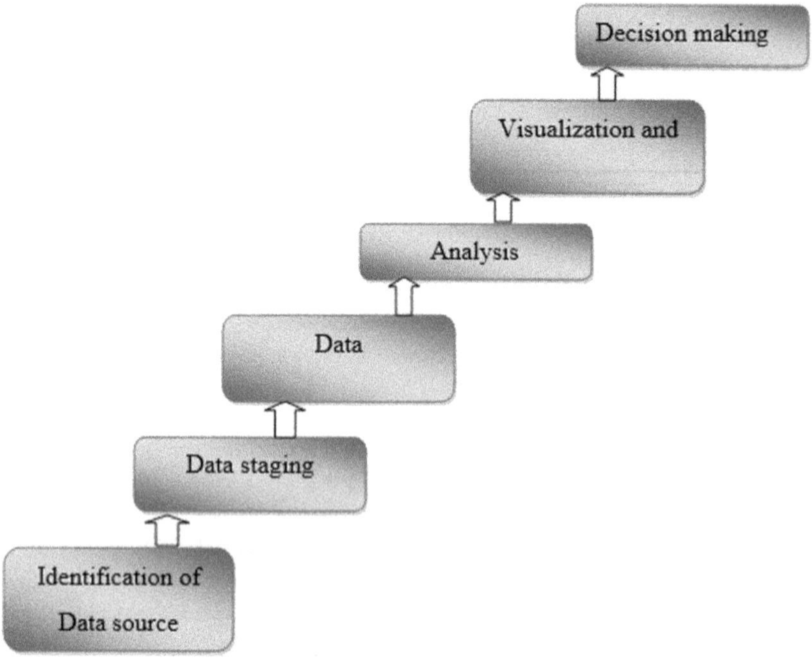

FIGURE 10.3 Business intelligence components (based on [17]).

information from enormous volumes of data to enable business analysts to make informed strategic decisions [13]. BI ensures quality service delivery, continued survival, and the success of an organization. Data analytics is a technically demanding area which majors on the programs and methods that work with the raw data from multiple heterogeneous sources to establish trends within the data, as well as statistical analysis [14–16]. The key differentiating factor is that BI revolves around business knowledge, whereas data analytics is data centered. The BI pipeline includes the following components as shown in Figure 10.3:

- Data source identification.
- Data staging.
- Data warehouse/mart.
- Data analysis.
- Visualization and reporting.
- Decision making.

10.4.1 Data Source Identification

The data are collated from different sources, operational systems, and external systems. Operating systems include pharmaceutical transaction processing systems, patient databases, medical health human resource management systems, and billing systems, while external sources include surveys, health reports, research publications, etc. The raw data in a patient database may include patient temperature, blood pressure, weight, whereas doctor's name, registration number, area of specialization, and years of service can be found in the human resource database, with data such as drug_name, drug_code, drug_description, and drug_quantity being present in the pharmaceutical database. The healthcare business analyst plays the major role in identifying or specifying relevant data sources to use in the BI system. Among the medical fraternity, the medical superintendent, doctors, and health officers are the key managerial staff viewed as the business analysts.

10.4.2 Data Staging

Data staging involves three key processes, namely extraction, transformation, and loading (ETL). This stage uses a blend of technologies in order to fulfill the ETL task. Data collected from multiple sources is temporarily stored in the staging area for cleansing and conversion into a format suitable for data analysis while pending uploading into a data warehouse or data mart. This stage is necessary since multiple sources vary in terms of standards and formats of data storage, which gives rise to inconsistencies. For example, a doctor's employee number may be stored differently in operational databases as D_ID or Doc_ID or Doc_Code, with each referring to the same doctor. Therefore, to avoid inconsistency and errors, there is a need to adopt a single standard format for ease of analysis.

10.4.3 Data Warehouse/Mart

A data warehouse is characterized by subject-oriented, integrated, non-volatile, and time-invariant data [17]. A data warehouse stores data by subject, not application. For example, hospital ordering is subject oriented in that it requires a host of data from various applications in order to arrive at a hospital ordering decision, such as when and how much consumables or disinfectants, etc., to order. The hospital stores application, patient application, and hospital human resource application provide the relevant data for order processing. The management responsible for hospital order processing requires this complete view from various tables from these different applications in order to arrive at an informed, cost-efficient and strategic decision. This enables consideration of all entities affected by the ordering decision.

A data warehouse is referred to as being integrated since the data comes from multiple sources, including structured, unstructured, and semi-structured data. A data warehouse contains time-invariant data in that it consists of historic data. "Non-volatile" implies that, once data are written, they cannot be changed. A data mart differs from a data warehouse in that it focuses on a single department, whereas a data warehouse can be perceived as a collection of data marts.

10.4.4 Data Analysis

Data analysis involves examination of two forms of data, namely dynamic data and static data. There are various tools which enable analysis of the historic and current data from a data warehouse/mart. Excel spreadsheets are the earliest form of BI tools used to analyze data. Some other tools are Online Analytical Processing (OLAP) tools, which enable the execution of complex multidimensional queries, e.g., Cognos. Online Transaction Processing (OLTP) deals with transaction-related data.

10.4.5 Visualization and Reporting

Analytical tools do not necessarily need to provide a pictorial representation; some representations are best depicted in report form, such as a cash flow statement for accounting management. Visual dashboards comprise of a mixture of pictorial and report representations that provide data mining detail in real time. This provides quick, visually captivating access to information by decision makers. Examples of such tools, with powerful and interactive data visualization dashboards, include Power-BI, Domo and IBM Cognos Analytic [18].

10.4.6 Decision Making

Decision making is the application of knowledge, acquired from data analytics by analysis tools, from the huge volumes of data in the data warehouse. Based on the relationships in the data, the organization can quickly identify trends and the nature of progress that leads to effective decision making. They can take advantage of this information to formulate strategic goals to increase business profit, improve service delivery, and achieve continued growth of the business. The knowledge can be used to make accurate predictions for investment decisions and categorization of clients. This provides increases in overall business efficiency and profitability.

10.5 BUSINESS INTELLIGENCE TOOLS

Business intelligence tools enable identification of interesting patterns or trends in the data. In turn, these provide the information needed to transform an organization into a successful one. The golden insights derived from the data facilitate information-driven strategic decision making. BI tools may also be viewed as business intelligence software applications, developed with the intention of retrieving, analyzing, transforming, and visualizing data to enable the continued existence, survival, and success of a business entity. These insights assist in organizational growth by providing actionable business information, solutions to seasonal problems of low business performance issues, and prediction of future outcomes and potential effects of introducing a new product.

There are several BI tools available, including SAP Business Intelligence, MicroStrategy, Datapine, SAS Business Intelligence, Yellowfin BI, QlikSense, Zoho Analytics, Sisense, Microsoft Power-BI, Looker, Clear Analytics, Tableau, Oracle BI, Domo and IBM Cognos Analytic [18].

10.5.1 Introduction to Power-BI

Microsoft Power-BI is a web-based, analytics software tool for business, which has exceptional data visualization capabilities. This means that end users can use a BI tool, regardless of their location. They can readily identify interesting patterns in real time because it is a web-based tool, which can be accessed from pretty much anywhere. This software application simplifies end-user integration of their apps, allows swift delivery of reports, and provides for dashboards in real time [19].

10.5.2 Results and Discussion

In this section, the UNCOVER COVID-10 challenge dataset from kaggle.com was analyzed, using the business intelligence tool, Microsoft Power-BI [19], to provide insights from the data. The UNCOVER COVID-10 challenge dataset contains data on coronavirus cases at different regional and international levels. The Roche Data Science Coalition (RDSC) established the UNCOVER COVID-10 public dataset. The dataset consists of huge volumes of raw data collected from various countries, geo-spatial data, and social distancing data [20]. Raw data are used for exploratory analysis by data analysts, whereas the exploratory research provides answers to questions such as:

- Who is at high risk of being infected by the virus?
- How many people have been infected?
- What is the predicted number of infections or deaths?
- At what rate is the virus spreading?
- Which countries or regions are affected?
- Which variables affect the spread?

The information derived from the exploratory data analysis assists in monitoring and controlling the spread of the virus. Medical experts, staff, medical policy makers, World Health Organization (WHO), and a host of other entities will use this information to prepare and rapidly respond to potential cases that may likely arise. The data analysts employ data analytics tools, such as Power-BI, to aid in exploring the data at varying levels (country, continent, and global levels) and can provide real-time visual representations on the dashboard. The dashboard provides a user-friendly, quick, and easy way of interpreting trends in data for better response and greater awareness. Figures 10.4a-e, provide a step-wise illustration on how the BI tool can be used to analyze the UNCOVER COVID-10 dataset. Figure 10.4f shows a simple graphical visualization of the new cases and new deaths recorded at an international level by WHO for the period from January to March 2020 in year format, whereas Figure 10.4g shows the same data in date-wise format. The highest number of new cases (39,114) was reported on 17 February 2020, with the corresponding highest number of deaths being reported on the same day, 212. Various visualization methods are available in Power-BI, such as a pie chart, map, bar chart, and 3D chart, to achieve greater insights. These visuals can readily be generated by selecting the appropriate choice from the visualization method. The section that follows (Section 10.5.2) provides a step-by-step guide on how to get started with Power-BI, how to work with data from multiple sources, and how to create a dashboard and share data.

10.5.2.1 Getting Started with Power-BI

1. Download the UNCOVER COVID-10 challenge dataset from https://www.kaggle.com/roche-data-science-coalition/uncover
2. Extract and save the dataset in a folder and rename it "my_dataset"
3. Install Power-BI http://app.powerbi.com
4. Download and run the Power-BI setup
5. Launch Power-BI as shown in Figure 10.4a
6. Click Get data -> select Text/CSV -> Connect button in Figure 10.4b
7. Select CSV file ...\my_dataset\UNCOVER\WHO in Figure 10.4c
8. Click Load dataset as shown in Figure 10.4d
9. Tick field, check box new_case and new_deaths under fields as shown in Figure 10.4e to achieve visualization (Figure 10.4g)
10. Click File->Save as ->filename.pbix extension, e.g., COVID_19.pbix to save in the location of your choice
11. Click Publish menu to share the report

10.5.2.2 Working with Multiple Data Sources

1. Follow steps 6, 7 and 8 in Subsection 10.5.2.1 to add data from multiple sources
2. Click Modeling-> Manage relations-> auto detect to create relationships
3. Click + sign to add more canvas pages to place separate visualizations without clustering on the same workspace
4. Click on any visualization chart of your choice to view cross filtering
5. Click Publish menu to overwrite previous report or to rename report

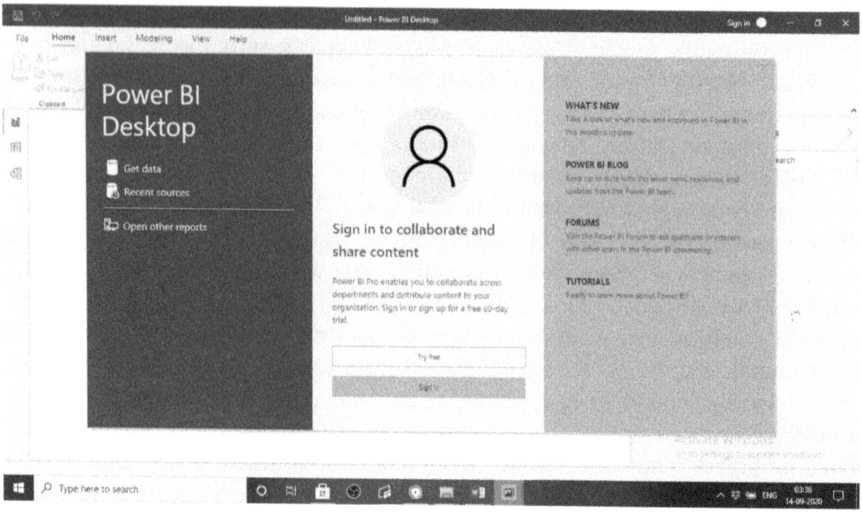

FIGURE 10.4 Getting started with Power BI. a: Power-BI launch; b: Get data; c: CSV file selection; d: Load dataset; e: Select fields; f: Data visualization in year, January to March 2020; g: Date-wise data visualization, January to March 2020.

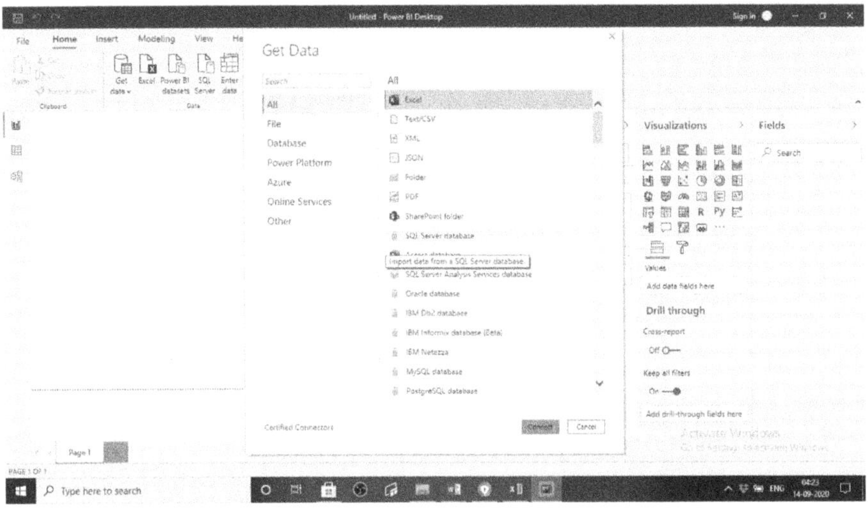

FIGURE 10.4 (Continued)

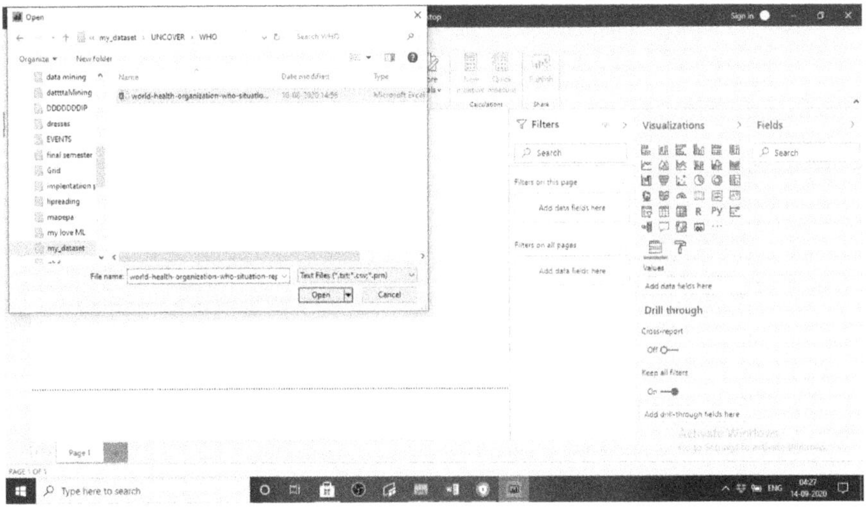

FIGURE 10.4 (Continued)

10.5.2.3 Creation and Sharing of Dashboard

1. Select pin icon to add visualizations to the new or existing dashboard
2. Rearrange visualization as desired
3. Click Share icon to view and share dashboard
4. Select corner of visualization to modify visuals

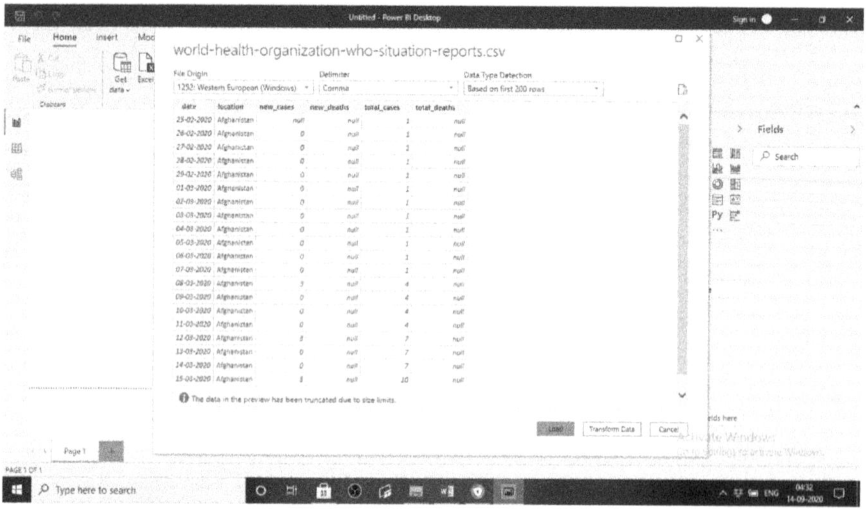

FIGURE 10.4 (Continued)

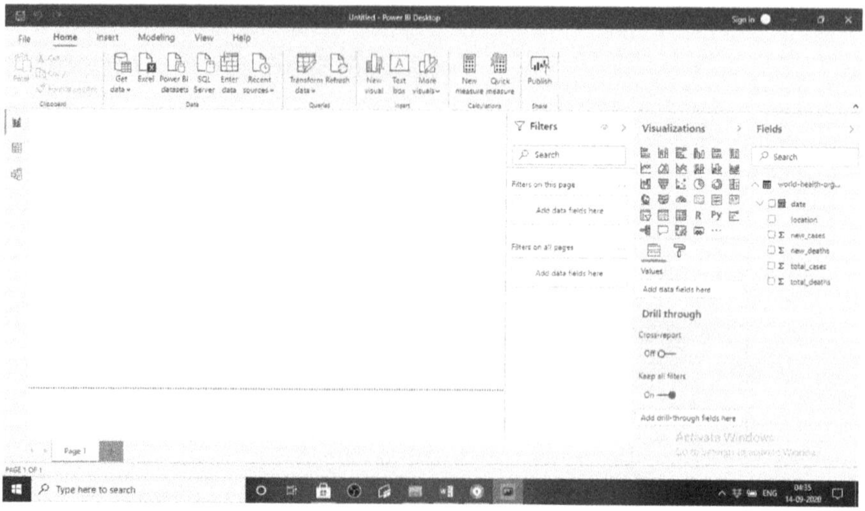

FIGURE 10.4 (Continued)

10.5.3 Recommender System

Recommender systems (RS) propose things or items of interest to an end user by luring users with relevant propositions, based on the choices they make. These recommender systems have experienced an exponential growth and development over the past few years. Amazon, YouTube, and Netflix are among the most commonly known RS for suggesting service or products to their clients. The same principle used in developing retail recommender systems can be applied to various fields to

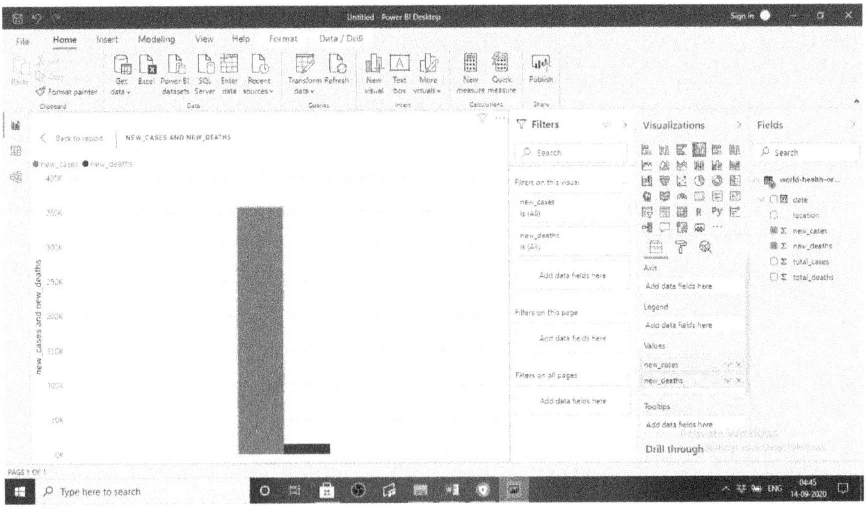

FIGURE 10.4 (Continued)

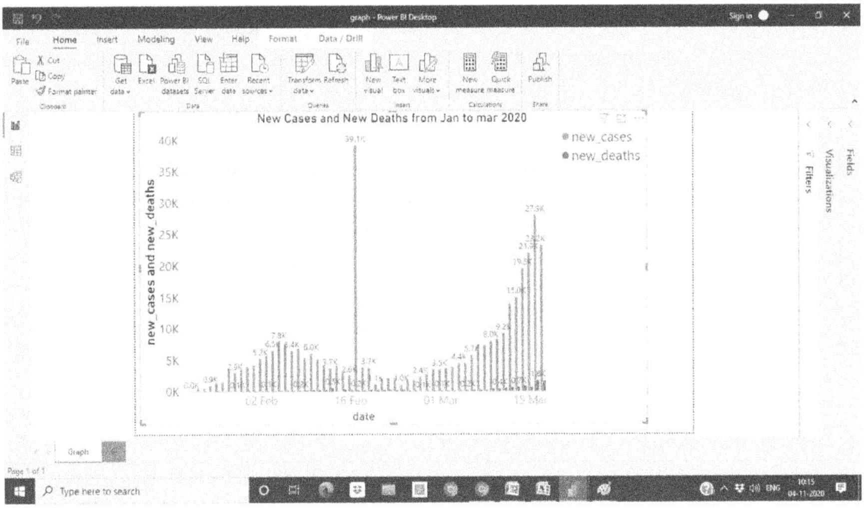

FIGURE 10.4 (Continued)

come up with domain-specific RS, such as personalized healthcare RS. A health recommender system (HRS) is a specialization of recommender systems [21]. In HRS, a recommendable item of interest is a piece of scientific generalization of medical information that is acceptable, which, in itself, is not linked to an individual's medical history. In a different perspective, health recommender system can be used for filtering data used to predict the inclination toward a particular grouping.

Over the past few years, vast amounts of data have been collected and generated in the medical field, such as laboratory results, treatment plans, and reports. Hence, the amount of digital information available for decision making is enormous but scattered across different sources. There has been an increasing global demand for health information [22]. As a solution, health recommender systems (HRS) are designed to help meet this need by providing suggestions on healthcare issues. HRS suggestions are data driven by decision-making systems, which recommend appropriate healthcare information for the benefit of the various actors in the medical and related fields, such as doctors, nurses, patients, and healthcare service providers.

10.6 FINDINGS FROM EHR RECORDS, USING MACHINE LEARNING ALGORITHMS

The healthcare recommender system is built using machine learning algorithms. EHR datasets provide the much-needed data to train, validate, and test machine learning-based recommender systems. The precision of the recommender system should be comparable to or above that of a human expert in the field in order to be of value. For example, if certain medical symptoms are inputted into an HRS, the output should be a correct predictor of illness and of a treatment plan or medication. These machine learning algorithms perform two main tasks, namely descriptive or predictive analytics.

10.6.1 Descriptive Analytics

Descriptive analytics characterize the general or typical properties of data in the data source [23]. They highlight values which can be viewed as noise or outliers. Descriptive analysis includes associations, clustering, and summarization. Association analysis is concerned with identifying frequent patterns, such as gene base sequences in genomic data analysis and protein amino acid sequences in proteomic analysis, so as to identify genetic disease or evolution. Clustering can be used to identify a group of people prone to a particular disease, such as diabetes, cancer, or malaria, and is based on various input factors, such as blood group, age, weight, and height, to mention but a few. Summarization involves picturing the data in various formats, such as distribution graphs, pie charts, curves, measures of dispersion, boxplots, scatter plots, and so on, to gain better insight or a general view of the nature of the data.

Algorithms

- Clustering analysis is an unsupervised form of learning in which there are no training datasets. The class labels are unknown. Clustering categorizes group data into clusters, using similarity indexes, by minimizing the intra-cluster distance and maximizing the inter-cluster distance. Examples of commonly used clustering algorithms include *K*-Means and *K*-Medoids.

- Association analysis seeks to find frequent patterns, subsets, or substructures within the dataset. The interpretation of the associations is dependent on the domain expert. Some examples of association analysis algorithms include Apriori, Elcat, and FP Growth.
- Outlier analysis detects anomalies in data. In a majority of cases, outliers are viewed as noisy data and are usually discarded. However, they can be used for anomaly detection. Outliers are easily or readily identified during data summarization in scatter plots, box plots, or cluster analysis.

10.6.2 Predictive Analytics and Insights

Predictive analytics perform inferences on the current data in order to make predictions [24]. The following forms under predictive analysis include classification, prediction, and time series analysis [25]. Classification analysis is the process of developing a model for describing data classes for the purpose of prediction of categorical data. Examples of categorical data are cancerous or non-cancerous cells, and healthy or unhealthy cells [26]. Prediction analysis models involve continuous value functions and are used to predict future values, like the spread of diseases in the human body or environment, such as COVID-19, cholera, tuberculosis, and typhoid.

Algorithms

- Classification is a form of supervised learning, being a non-numerical decision analysis based on decision/class labels, and is predictive, predicting class labels or groupings. Examples of such algorithms include Decision Tree, Naïve Bayes, Neural Networks, Support Vector Machine, and genetic algorithms.
- Regression is supervised learning, with numerical decision analysis based on the decision/class label and is also predictive. It forms an equation that fits certain points known as the line-of-best-fit. Examples of such algorithms include Regression Tree, Logistic Regression, and Multi Linear Regression.

10.7 CONCLUSION

Healthcare data form the basis on which the numerous healthcare actors, namely patients, doctors, nurses, researchers, etc., provide a high-quality healthcare service. Over the past few years, there has been an exponential growth in both healthcare demands and the data generated by the medical fraternity. Advances in sensor technology, connectivity, and infrastructure have been the major drivers of huge volumes of data, termed big data. Big data cannot be readily handled by traditional methods, but requires specialized data analytics technologies in order to draw useful information from the data. This chapter provides a practical guide on how the Power-BI business intelligence tool of Microsoft is used to gain useful insights from EHR, and the UNCOVER COVID-10 challenge dataset. This provides hands-on understanding of the key components involved in data analytics, namely data collection, data cleansing, data warehousing, descriptive/predictive analysis, visual presentation/reports, and decision making.

A detailed description of a host of techniques available at each stage of data analytics is provided in this chapter to ensure successful data processing. An insight of how recommender systems are built, based on descriptive and predictive analysis using machine learning algorithms, is highlighted. The recommender systems enable timely, cost-effective, efficient, profitable, and life-saving decisions to be made to meet the increasingly highly personalized, dynamic patient demands in a user-friendly way. The satisfaction of patients is the goal of all medical health service providers, contributing to the health and wellbeing of the entire human population.

REFERENCES

1. Kumar, Y., Sood, K., Kaul, S., & Vasuja, R. (2020). Big data analytics and its benefits in healthcare. In Kulkarni, A. (ed.), *Big Data Analytics in Healthcare* (pp. 3–21). Springer, Cham.
2. De Mauro, A., Greco, M., & Grimaldi, M. (2016). A formal definition of Big Data based on its essential features. *Library Review.*
3. Dash, S., Shakyawar, S. K., Sharma, M., & Kaushik, S. (2019). Big data in healthcare: management, analysis and future prospects. *Journal of Big Data, 6* (1), 54.
4. Tang, P. C., Ash, J. S., Bates, D. W., Overhage, J. M., & Sands, D. Z. (2006). Personal health records: definitions, benefits, and strategies for overcoming barriers to adoption. *Journal of the American Medical Informatics Association JAMIA, 13* (2), 121–126.
5. Agrawal, R., & Prabakaran, S. (2020). Big data in digital healthcare: lessons learnt and recommendations for general practice. *Heredity, 124*, 1–10.
6. Li, J. S., Zhang, Y. F., Tian, Y., Soto, S. V., Luna, J., & Cano, A. (2016). Medical big data analysis in hospital information system. In Soto, S. V., Luna, J., & Cano, A. (eds.), *Big Data on Real-World Applications.* IntechOpen Limited, 65.
7. Fiaidhi, J. (2020). Envisioning insight-driven learning based on thick data analytics with focus on healthcare. *IEEE Access, 8*, 114998–115004.
8. Carter, J. Y., Lema, O. E., Wangai, M. W., Munafu, C. G., Rees, P. H., & Nyamongo, J. A. (2012). Laboratory testing improves diagnosis and treatment outcomes in primary health care facilities. *African Journal of Laboratory Medicine, 1* (1), 8.
9. Chen, M., Hao, Y., Hwang, K., Wang, L., & Wang, L. (2017). Disease prediction by machine learning over big data from healthcare communities. *IEEE Access, 5*, 8869–8879.
10. https://www.karexpert.com/dhs/smart-doctor-e-prescription.
11. Han, J., Kamber, M., & Pei, J. (2011). Data mining concepts and techniques third edition. In *The Morgan Kaufmann Series in Data Management Systems*, 83–124.
12. Dunham, M. H. (2006). *Data Mining: Introductory and Advanced Topics.* Pearson Education India.
13. Abusager, K., Baldwin, M., & Hsu, V., 2020. Using power BI to inform clostridioides difficile ordering practices at an acute care hospital in central Florida. *American Journal of Infection Control, 48* (8), pp. S57–S58.
14. Raghupathi, W., & Raghupathi, V. (2014). Big data analytics in healthcare: Promise and potential. *Health information science and systems, 2* (1), 3.
15. Yin, H., Sun, Y., Cui, B., Hu, Z., & Chen, L. (2013, August). LCARS: a location-content-aware recommender system. In *Proceedings of the 19th ACM SIGKDD International Conference on Knowledge Discovery and Data Mining* (pp. 221–229).

16. Wu, P. Y., Cheng, C. W., Kaddi, C. D., Venugopalan, J., Hoffman, R., & Wang, M. D. (2016). –Omic and electronic health record big data analytics for precision medicine. *IEEE Transactions on Bio-medical Engineering*, *64* (2), 263–273.
17. Theraja, Reema. 2011. "Datawarehousing". Oxford University Press.
18. https://mopinion.com/business-intelligence-bi-tools-overview/
19. www.powerbi.microsoft.com
20. Siddique Latif, M. U., Manzoor, S., Iqbal, W., Qadir, J., Tyson, G., Castro, I. & Crowcrroft, J. (2020). *Leveraging Data Science to Combat Covid-19: A Comprehensive Review*, TechRxiv, Preprint, 1–19.
21. Ricci, F., Rokach, L., & Shapira, B. (2011). Introduction to recommender systems handbook. In Ricci, F., Rokach, L., Shapira, B., Kantor, P. B. (eds.), *Recommender Systems Handbook* (pp. 1–35). Springer, Boston, MA.
22. Gavgani, V. Z. (2010, November). Health information need and seeking behavior of patients in developing countries' context; an Iranian experience. In *Proceedings of the 1st ACM International Health Informatics Symposium* (pp. 575–579).
23. Hand, D., Manila, H., and Smyth, P. (2001). *Principles of Data Mining.* MIT Press, Cambrdige, MA.
24. Venkatesh, R., Balasubramanian, C., & Kaliappan, M. (2019). Development of big data predictive analytics model for disease prediction using machine learning technique. *Journal of Medical Systems*, *43* (8), 272.
25. Witten, I. H., & Frank, E. (2002). Data mining: practical machine learning tools and techniques with Java implementations. *Acm Sigmod Record*, *31*(1), 76–77.
26. Muhammad, K., Khan, S., Del Ser, J., & de Albuquerque, V. H. C. (2020). Deep learning for multigrade brain tumor classification in smart healthcare systems: A prospective survey. *IEEE Transactions on Neural Networks and Learning Systems,* 1–16.

11 Machine Learning-Based Data Classification Techniques in Healthcare Using Massive Online Analysis Framework

B. Ida Seraphim and E. Poovammal

CONTENTS

11.1 INTRODUCTION

The massive growth in the use of technology in the healthcare industry has brought a huge and welcome change. Patients can have access to new cutting-edge technological treatments, that are less painful and result in quicker healing. The patients can have a remote consultation with specialists, and targeted treatments, in addition to normal treatments. The available apps can lead to improved patient healthcare monitoring. The new treatment technology gives better outcomes, which enhances the quality of the patient's life. The use of navigation apps, movie or music recommendations, purchase and order tracking suggestions for online purchases has made machine learning an integral part of everyone's life. Machine learning in the field of healthcare is at the early stages of development. Figure 11.1 shows various areas in health care where machine learning is used.

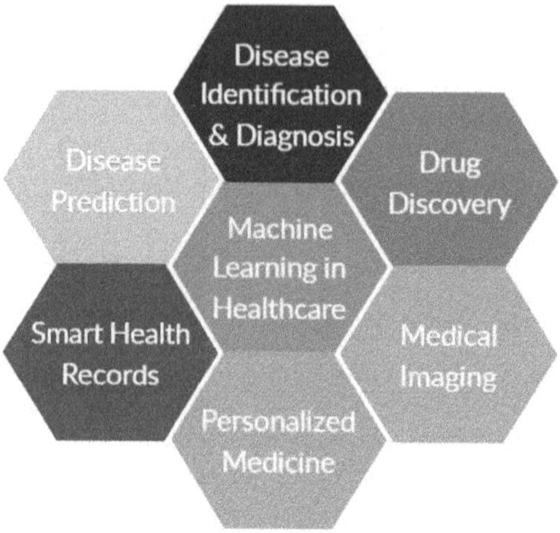

FIGURE 11.1 Machine learning – health care.

11.1.1 Disease Identification and Diagnosis

Achieving accurate disease identification and diagnoses is the ultimate challenge for the global healthcare system. It is estimated that 5% of patients receive an incorrect diagnosis each year. These errors are common in patients with serious health conditions [1]. Machine learning plays a key role in detecting the disease, monitoring health conditions, and recommending measures to prevent the disease, as manual diagnosis and prediction of disease is very difficult [2]. These improvements from machine learning can range from minor conditions to serious diseases, such as cancer, that are very difficult to detect at the early stage.

11.1.2 Drug Discovery

Research and development departments of the pharmaceutical industry are very keen on finding therapies for complex diseases. It is a costly and time-consuming process to discover and manufacture a new drug [2]. The newly discovered drug has to undergo a number of clinical trials to ensure the effectiveness of the drug against the disease in question and its safety with respect to the patient [3]. Drug discovery is divided into four phases. Phase 0 encompasses basic research and preclinical tests. Phase 1 encompasses a study on the side effects, toxicity, and on the drug's pharmacokinetic dose-response relationship. Phase 2 determines drug performance. Phase 3 compares the safety and effectiveness of the new drug with those of current medications available for the condition in question. Phase 4 is optional; it takes place after the drug has received approval from the regulatory bodies, to find the drug's long-term effectiveness and side effects [3]. Machine learning algorithms help in simulating the drug discovery and manufacturing processes, thereby reducing the costs involved.

11.1.3 Medical Imaging

Machine learning (ML) in medical imaging helps to find microscopic anomalies in the scanned images. ML helps the doctor to diagnose the disease properly and treat the patients more efficiently. Traditionally, X-rays or computed tomography (CT) scans are used to diagnose a disease. But with increasing numbers of different kinds of diseases, the traditional methods are insufficient to achieve effective diagnosis [2]. ML technology is used to build innovative tools for prediction. For example, 3D radiological imaging uses ML technology.

11.1.4 Personalized Medicine

Hospitals maintain Electronic Health Records (EHR) used by doctors to provide personalized treatments to the patients, depending on their needs. ML technology is used to forecast disease risk and to reduce the amount of the doctor's time spent on disease treatment decisions. For example, the IBM Watson healthcare system uses

ML technology. IBM has developed the Watson healthcare system to address health applications in oncology, genomics, etc. [4]. Many start-up companies have started to address all possible aspects of healthcare applications of ML.

11.1.5 Smart Health Records

Trending technologies reduce the time spent on data entry work. It is time consuming to maintain up-to-date information on a patient's health history daily [2]. Connected health is a smart health technology in which smartphones are connected together with medical wearable devices (glucometers, smartwatches, blood pressure monitors) *via* wireless technologies [4]. The patient's medical data are collected and updated daily, using smart health technology, achieved through ML techniques, that save time, effort, and cost in maintaining the patient's information.

11.1.6 Disease Prediction

ML technology is used to monitor and predict numerous diseases. Researchers collect a huge volume of patient data through various sources like satellites, social media, websites, etc. ML techniques are used to predict everything from insignificant diseases to serious chronic diseases [2]. For example, researchers at Nottingham University in UK [2] uses ML technology to create a system that scans patients' health records and predicts how many people will have a heart attack within 10 years. When compared with the traditional prediction system, a system that uses ML technology predicts more accurately.

11.2 TYPES OF HEALTHCARE DATA

The Indian healthcare system is one of the fastest-growing industries in the world. Healthcare analysis and predictions are made on various diseases like cancer, strokes, diabetes, and autism, using ML techniques. There are various types of cancer, but the ones with the highest frequencies are breast cancer, lung cancer, mouth cancer, etc. All these types of diseases generate different kinds of healthcare data. Healthcare data is broadly classified into three types (Figure 11.2).

11.2.1 Clinical Data

Clinical data are collected during the time of treatment of the patients. Figure 11.2 shows that clinical data are broadly classified into six categories. EHR is the most direct form of clinical data which is not provided for the benefit of outside researchers. EHR data are collected from organizational reports, demographic statistics, diagnoses, treatments, drug prescriptions, laboratory tests reports, physiological monitoring data, hospitalization data, patient insurance reports, etc. [5]. Administrative data are most probably related to HER, like hospital discharge data reported to government agencies. Ailment registries often provide serious and sensitive information

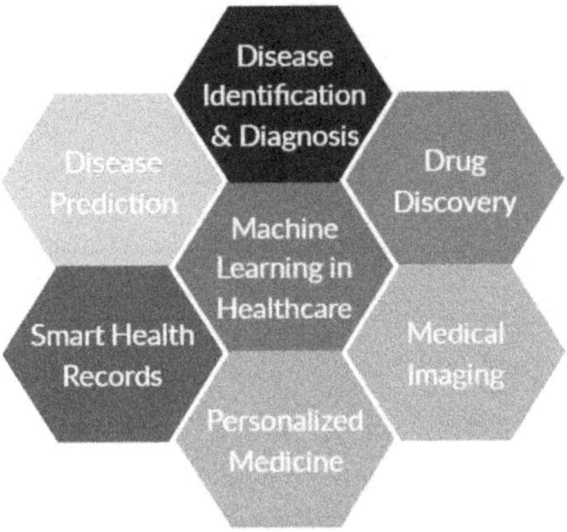

FIGURE 11.2 Types of healthcare data.

with respect to the handling of patient conditions [5]. Ailment registers are clinical information systems that track key data for chronic conditions such as autism, Alzheimer's, asthma, cancer, heart disease, and diabetes. Claims data are the billable interactions (insurance claims) between insured patients and the healthcare delivery system. The claims data are widely classified as either in-patient, pharmacy, outpatient, or enrollment [5]. Health surveys are mainly conducted to evaluate population health accurately, whereas national surveys are conducted to determine disease prevalence estimates.

11.2.2 Sensor Data

Sensor data are continuous time-series data that are generated in an ordered sequence. These data are mainly generated from wearable devices and wireless monitoring devices. Wearable health monitoring devices include microsensors, wearable devices, and implants [6]. These sensors re called biosensors, and help to monitor physiological variables like heart rate, blood pressure, oxygen saturation, electrocardiogram, etc. A wearable device integrates mobile healthcare into our daily routine.

11.2.3 Omics Data

Omics data is a huge collection of complex, high-dimensional data, such as genomic data, transcriptomic data, and proteomic data. Genomic data includes DNA sequence, used in the bioinformatics field. Transcriptomic data deals with transcriptional control of gene expression, in the form of the abundance of multiple mRNA transcripts taken from biological samples [7]. These samples are analyzed to extract the features from a dataset. Proteomic data deals with the proteins synthesized in cells, tissues, organs, and organisms. It indicates the functional translational and post-translational regulation of gene expression in the cell, tissue, etc.

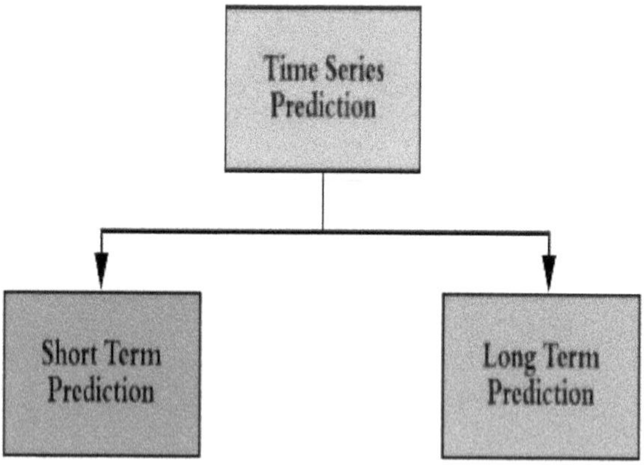

FIGURE 11.3 Time-series prediction categories.

11.3 TIME-SERIES DATA IN HEALTHCARE

The changes exhibited by the clinical variables at different time points is called time-series data. In the medical field, time-series data forecasting is applied to predict disease advancement, death rate, and time-related risk associated with the disease [8]. In time-series data, the features keep changing, depending on the time frame. The changes often represent important information. For example, blood glucose level and blood pressure monitored over time are considered to be important health indicators [9]. Time-series prediction is divided into two categories: short-term and long-term prediction. Short-term analysis requires regression analysis to precisely predict the future outcome. With long-term analysis, data trends are observed, and the future outcome is predicted. Figure 11.3 shows the different categories of time-series predictions.

11.3.1 Time-Series Analysis

Time-series analysis is a collection of specialized methods that illustrates the trends in the data over time. It incorporates past error information and observations, from which it predicts the estimated future outcome.

11.4 MACHINE LEARNING ALGORITHMS ON CLASSIFICATION TASKS

ML trains the machines to identify and discover different patterns from data from medical sources. It uses ML algorithms to predict its outcome. ML techniques are broadly classified into supervised learning, unsupervised learning, semi-supervised learning and reinforcement learning. In supervised learning, the classifiers are trained on the labeled instances to predict the unlabeled instances. The classification model is an example of supervised learning techniques. Unsupervised learning is used to train the unlabeled instances. The clustering model is an example of unsupervised learning. Semi-supervised learning combines small amount of labelled data with unlabeled data for prediction. In reinforcement learning, a domain expert labels the unlabeled data [10].

11.4.1 Classification

Classification is the supervised learning approach where the machine learns from the given data and makes new predictions. It is called supervised learning because the labels indicate the class to which it belongs in the training data. When the new data or example comes in, it is classified based on the training set [11]. Heart disease detection is one example of a classification problem, which states whether a person has heart disease or not.

11.4.2 Classification Model

The classification model predicts the conclusion and assigns the label to the input instances given for training. The four different ML classification tasks are:

1. Binary classification
2. Multi-class classification
3. Multi-label classification
4. Imbalanced classification

11.4.3 Binary Classification

Binary classification is when the input instances are divided into two classes or two class labels [11]. The outcome of binary classification can be yes or no, true or false, etc. Examples of binary classification include:

- Heart disease detection (yes or no)
- COVID-19 (positive or negative)
- Cancer detection (yes or no)

In clinical binary classification, the classes are divided into normal instances and abnormal instances. For example, in cancer detection, cancer detected is an abnormal class, and no cancer detected is a normal class. The class label 0 is assigned to the normal class and class 1 to the abnormal class. *K*-Nearest Neighbor, Decision Tree, Support Vector Machine, and Naïve Bayes are some of the widely used binary classification algorithms.

11.4.4 Multi-Class Classification

The multi-class classification task involves more than two class labels. Unlike binary classification, multi-class classification is classified as among the range of known classes. The number of class labels may differ from one problem to another [11]. An example of multi-class classification is the classification of cancerous tumors as normal, benign, and malignant. Other examples include:

- Face classification
- Plant species classification

Multi-class classification covers the events that have a categorical outcome. Multi-class classification can use the algorithms that are used for binary classification. *K*-Nearest Neighbor, Decision Tree, Naïve Bayes, and Gradient Boosting are some of the algorithms used widely in multi-class classification.

11.4.5 Multi-Label Classification

A multi-label classification task involves two or more class labels, where they are used to predict one or more class. For example, in photo identification, there are multiple objects in a scene, such as a person, an apple or books, etc. The classification model is used to predict multiple known objects in a photo at a given point in time [11]. The single label is used to predict each example in binary and multi-class classifications. Thus, the multi-label classification model predicts multiple outputs.

Multi-label classification cannot use the binary and multi-class classification algorithms directly. Specialized versions of traditional classification algorithms are used in the case of multi-label classification. Some of the specialized versions of multi-label classification algorithms are multi-label Decision Trees, multi-label Random Forest, and multi-label Gradient Boosting.

11.4.6 Imbalanced Classification

In an imbalanced classification, the number of instances is not equally distributed. The larger number of instances in the training dataset fits the normal class, with a smaller number of instances belonging to the anomaly class [11]. Examples of an imbalanced classification are:

- Fraud detection
- Medical diagnostics test analysis
- Outlier analysis

The above-stated problems are modeled as binary classifications that require specialized techniques due to the imbalanced instances in the labeled classes. Carrying out under-sampling or over-sampling techniques in the training dataset changes the composition of the examples. The cost-sensitive techniques are used to make the final predictions. Some examples of imbalanced classification algorithms are cost-sensitive Decision Tree and cost-sensitive Support Vector Machine algorithm.

11.5 MASSIVE ONLINE ANALYSIS (MOA) FRAMEWORK FOR TIME-SERIES DATA

Massive Online Analysis (MOA) is a popular framework for both the non-streaming and streaming environments. MOA is an open-source freely available Java-based framework that includes several ML algorithms and visualization tools. MOA handles a massive amount of data from the incoming data streams within a short amount of time and with limited memory.

11.5.1 Setting the Environment

The latest version of MOA is downloaded from https://moa.cs.waikato.ac.nz. The compressed file contains the moa.jar file, an executable java jar file used by a Java application, or a command-line environment. The file also contains bin\moa.bat file for Windows and bin\moa.sh for Linux and Mac systems, through which the MOA uses the graphical user interface (GUI) [12]. Figure 11.4 shows the GUI of MOA.

Once the GUI is ready, click ‘Configure’ to set up the task and click ‘Run’ to launch the desired job. MOA runs more than one task concurrently. The GUI setup consists of four main steps, namely:

- Task selection
- Learner/algorithm selection

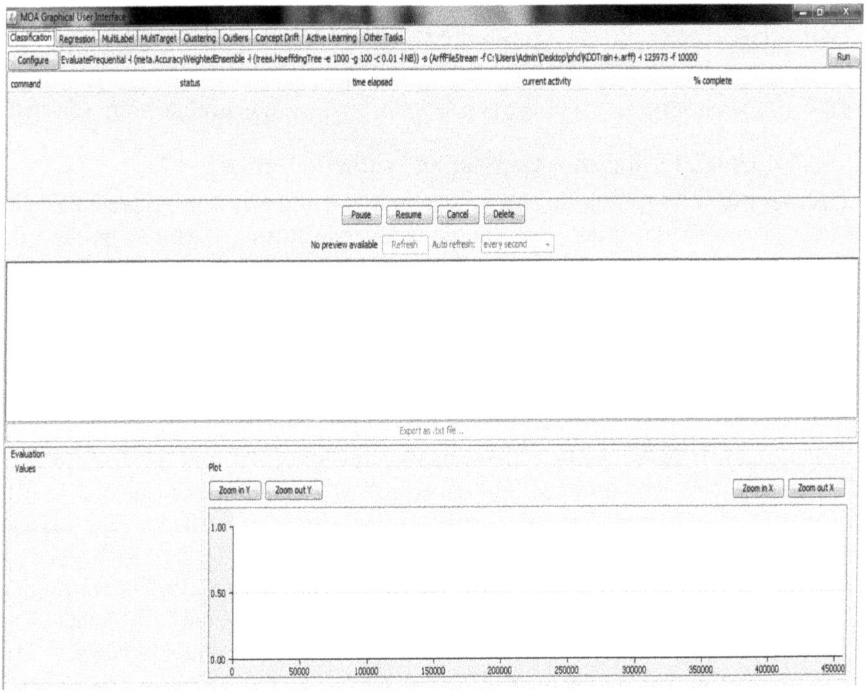

FIGURE 11.4 Massive Online Analysis GUI.

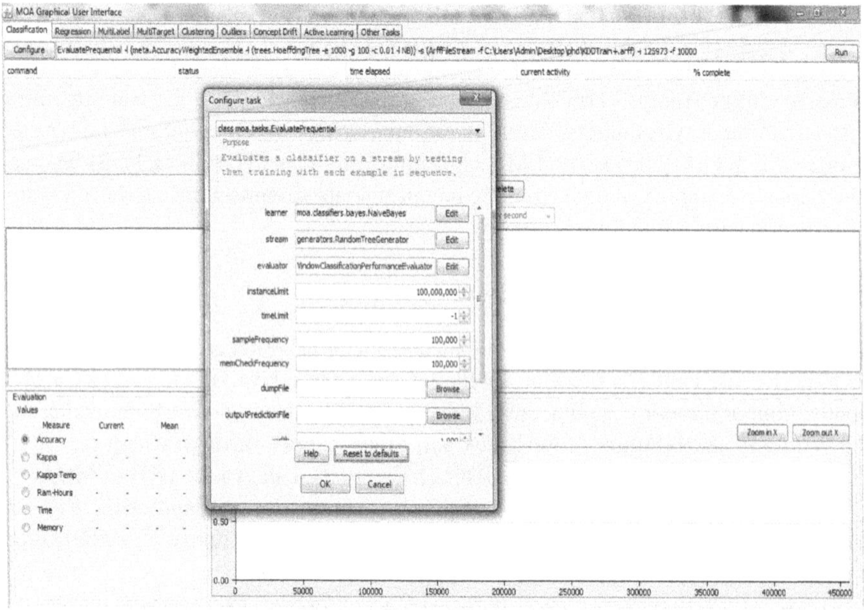

FIGURE 11.5 Synthetic data generator in MOA framework.

- Stream generator selection
- Evaluator selection

Figure 11.5 shows what kind of learner, steam generator, and evaluator are selected.

11.5.1.1 Upload Dataset or Generating Synthetic Dataset

MOA uses two ways of handling data execution. The first way is to generate the synthetic dataset with the available stream generator algorithms in the MOA framework. The second way is to upload the available benchmark datasets. Figure 11.5 shows the first way of generating a synthetic dataset from the built-in algorithms available in the framework. In Figure 11.5, the stream generator algorithm selected is Random Tree Generator, that creates a synthetic dataset with the in-built set of attributes. Figure 11.6 shows the second way of uploading the benchmark dataset and generating the stream.

The dataset is then uploaded, the instance limit specified, and the frequency of samples specified. The instance limit determines the number of instances present in the dataset. The frequency of samples shows how many instances are present between the samples of the learning performance. The default time limits, −1, says that there is no time limit to test or train instances. The dataset available in Kaggle UCI, and other repositories is in the CSV (comma-separated values) file format. The MOA framework accepts only .arff (Attribute-Relation File Format) file format. The file conversion of the .csv file to .arff file is made.

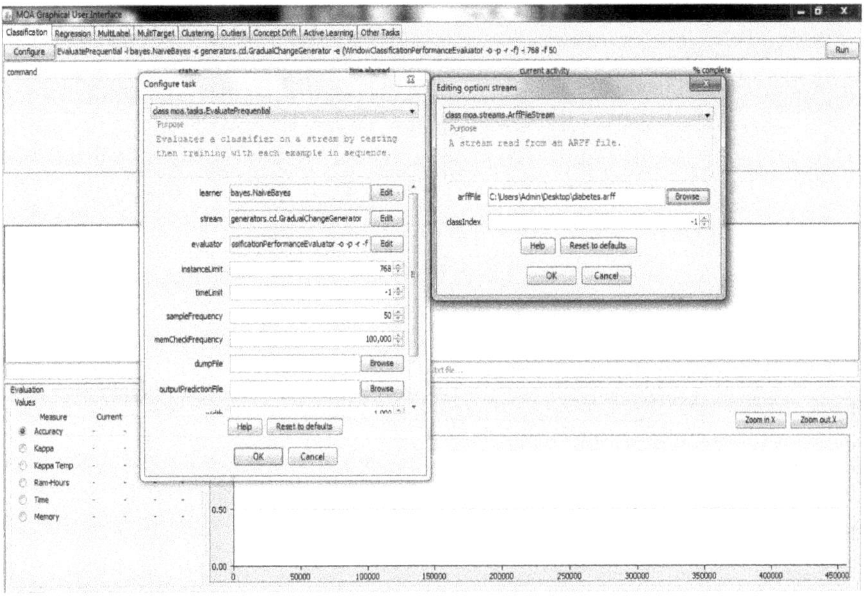

FIGURE 11.6 Uploading a dataset in the MOA framework.

11.5.1.2 Converting CSV File to .arff File Format

The benchmark dataset taken from the repositories is made available in .csv file format. The .csv file is converted to .arff file format because the MOA framework works only with .arff files. The WEKA tool converts the file from .csv to .arff. WEKA is also a free open-source framework like MOA. The University of Waikato, New Zealand, developed WEKA and MOA to perform data mining and ML tasks.

The WEKA tool is downloaded from the https://www.cs.waikato.ac.nz/~ml/weka/ link. Upon opening the installation and the WEKA tool, click on 'Tools', and select .arff viewer. Figure 11.7 shows the conversion from .csv file to .arff file made using WEKA.

The .arff viewer is opened, the file open is selected, and the .csv file to be converted to the .arff file is selected. Figure 11.8 shows how to load the .csv file into WEKA.

After the .csv file is loaded, the file is saved as .arff, and the .csv file is converted into the .arff file. Figure 11.9 shows how the .csv file is saved as an .arff file.

Now, the converted file is ready to be loaded into the MOA framework. The dataset is loaded into the MOA framework by selecting class moa.streams.Arfffilestream from the stream option in the configures task window.

MOA can run several tasks concurrently and compare the results. Click on the list of tasks, one by one, to see the performance of each task. MOA displays the result at the bottom of the GUI. It compares two different tasks. A red color represents the current task, and a blue color represents the previous task.

FIGURE 11.7 WEKA framework conversion from .csv file to .arff file.

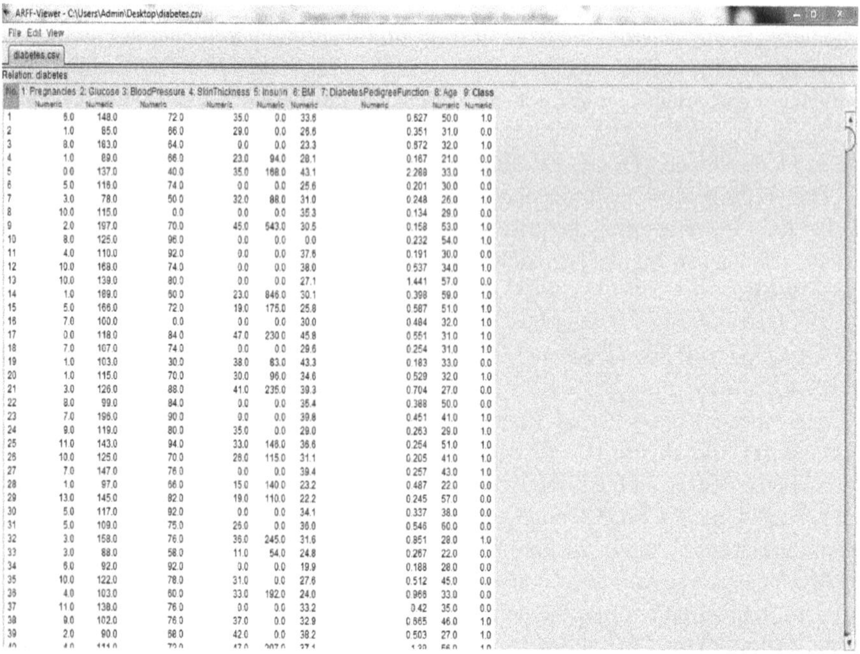
ARFF-Viewer - C:\Users\Admin\Desktop\diabetes.csv

File Edit View

diabetes.csv

Relation: diabetes

No.	1: Pregnancies Numeric	2: Glucose Numeric	3: BloodPressure Numeric	4: SkinThickness Numeric	5: Insulin Numeric	6: BMI Numeric	7: DiabetesPedigreeFunction Numeric	8: Age Numeric	9: Class Numeric
1	6.0	148.0	72.0	35.0	0.0	33.6	0.627	50.0	1.0
2	1.0	85.0	66.0	29.0	0.0	26.6	0.351	31.0	0.0
3	8.0	183.0	64.0	0.0	0.0	23.3	0.672	32.0	1.0
4	1.0	89.0	66.0	23.0	94.0	28.1	0.167	21.0	0.0
5	0.0	137.0	40.0	35.0	168.0	43.1	2.288	33.0	1.0
6	5.0	116.0	74.0	0.0	0.0	25.6	0.201	30.0	0.0
7	3.0	78.0	50.0	32.0	88.0	31.0	0.248	26.0	1.0
8	10.0	115.0	0.0	0.0	0.0	35.3	0.134	29.0	0.0
9	2.0	197.0	70.0	45.0	543.0	30.5	0.158	53.0	1.0
10	8.0	125.0	96.0	0.0	0.0	0.0	0.232	54.0	1.0
11	4.0	110.0	92.0	0.0	0.0	37.6	0.191	30.0	0.0
12	10.0	168.0	74.0	0.0	0.0	38.0	0.537	34.0	1.0
13	10.0	139.0	80.0	0.0	0.0	27.1	1.441	57.0	0.0
14	1.0	189.0	60.0	23.0	846.0	30.1	0.398	59.0	1.0
15	5.0	166.0	72.0	19.0	175.0	25.8	0.587	51.0	1.0
16	7.0	100.0	0.0	0.0	0.0	30.0	0.484	32.0	1.0
17	0.0	118.0	84.0	47.0	230.0	45.8	0.551	31.0	1.0
18	7.0	107.0	74.0	0.0	0.0	29.6	0.254	31.0	1.0
19	1.0	103.0	30.0	38.0	83.0	43.3	0.183	33.0	0.0
20	1.0	115.0	70.0	30.0	96.0	34.6	0.529	32.0	1.0
21	3.0	126.0	88.0	41.0	235.0	39.3	0.704	27.0	0.0
22	8.0	99.0	84.0	0.0	0.0	35.4	0.388	50.0	0.0
23	7.0	196.0	90.0	0.0	0.0	39.8	0.451	41.0	1.0
24	9.0	119.0	80.0	35.0	0.0	29.0	0.263	29.0	1.0
25	11.0	143.0	94.0	33.0	146.0	36.6	0.254	51.0	1.0
26	10.0	125.0	70.0	26.0	115.0	31.1	0.205	41.0	1.0
27	7.0	147.0	76.0	0.0	0.0	39.4	0.257	43.0	1.0
28	1.0	97.0	66.0	15.0	140.0	23.2	0.487	22.0	0.0
29	13.0	145.0	82.0	19.0	110.0	22.2	0.245	57.0	0.0
30	5.0	117.0	92.0	0.0	0.0	34.1	0.337	38.0	0.0
31	5.0	109.0	75.0	26.0	0.0	36.0	0.546	60.0	0.0
32	3.0	158.0	76.0	36.0	245.0	31.6	0.851	28.0	1.0
33	3.0	88.0	58.0	11.0	54.0	24.8	0.267	22.0	0.0
34	6.0	92.0	92.0	0.0	0.0	19.9	0.188	28.0	0.0
35	10.0	122.0	78.0	31.0	0.0	27.6	0.512	45.0	0.0
36	4.0	103.0	60.0	33.0	192.0	24.0	0.966	33.0	0.0
37	11.0	138.0	76.0	0.0	0.0	33.2	0.42	35.0	0.0
38	9.0	102.0	76.0	37.0	0.0	32.9	0.665	46.0	1.0
39	2.0	90.0	68.0	42.0	0.0	38.2	0.503	27.0	1.0

FIGURE 11.8 Loading .csv file to WEKA framework.

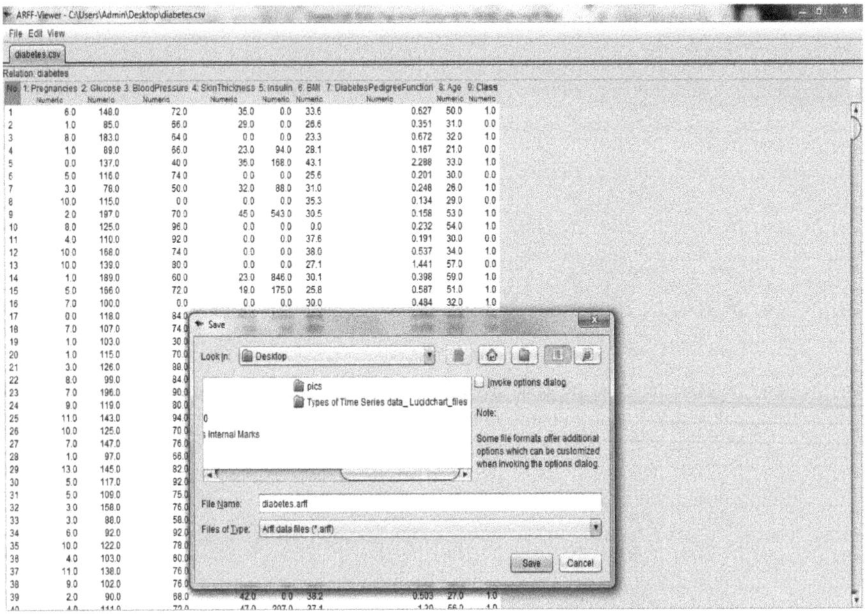

FIGURE 11.9 Saving as an .arff file.

For example, consider two different classifiers, namely Hoeffding Tree and Naïve Bayes, using prequential evaluation with 1,000,000 instances, and use a default stream generator called Random Tree Generator. Figure 11.10 shows the comparisons of the two different classifiers comparisons in the GUI of MOA.

EvaluatePrequential –l bayes.NaiveBayes –i 1000000 –f 1000
EvaluatePrequential –l trees.HoeffdingTree –i 1000000 –f 1000

11.5.2 Data Generator

The publicly available real-world benchmark datasets are significantly fewer. The well-known repositories, like UCI and Kaggle, contain many datasets, although not all the available datasets are suitable for streaming. Some researchers use their private data for research that cannot be reproduced by others. The MOA site itself provides four different datasets that are specifically available for this purpose. The researchers are mainly generating synthetic datasets by using synthetic stream data generator algorithms to show the results. The MOA tool has a number of stream generator algorithms to generate the synthetic data source. Table 11.1 shows the inbuilt synthetic data generators to create artificial data streams. The researchers can create their own synthetic data source generator algorithm, as MOA is an open-source Java-based tool.

Figure 11.11 shows the various stream generator algorithms available in MOA. After clicking 'Configure', the configure task window appears, then click on the

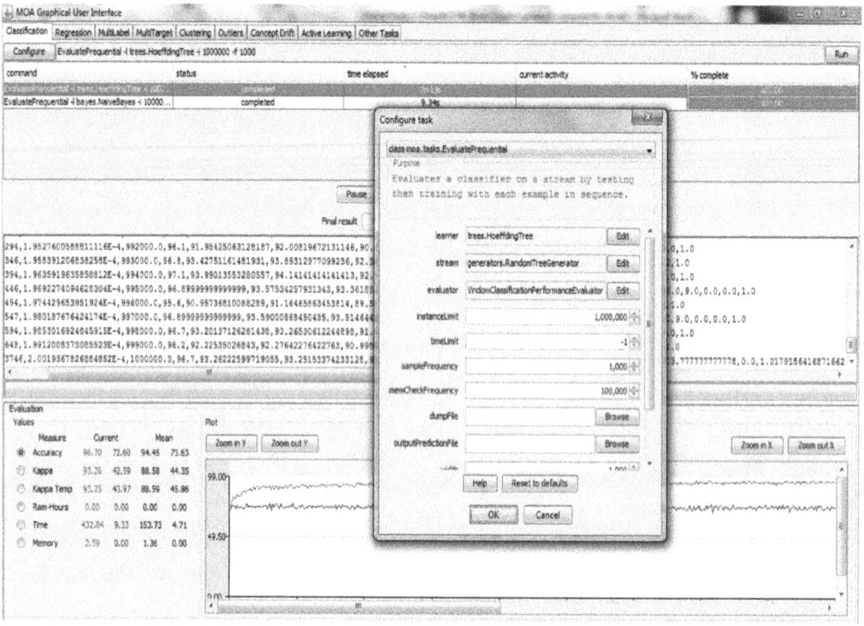

FIGURE 11.10 Comparison of two different classifiers in MOA GUI.

TABLE 11.1
Inbuilt Data Source Generators

	Attributes		
Data Source Name	**Nominal**	**Numeric**	**Class**
RTS/RTSN	10	10	2
RTC/RTCN	50	50	2
RRBFS	0	10	2
RRBFC	0	50	2
LED	24	0	10
WAVE21	0	21	3
WAVE40	0	40	3
GENF1-F10	6	3	2

'Edit' button beside the stream. Editing options: stream, window opens where all the stream generator algorithms appear.

11.5.3 Learning Algorithms

MOA is an excellent tool, with the latest, inbuilt ML algorithms. This tool has various tabs, such as classification, clustering, regression, and outlier detection. Each ML

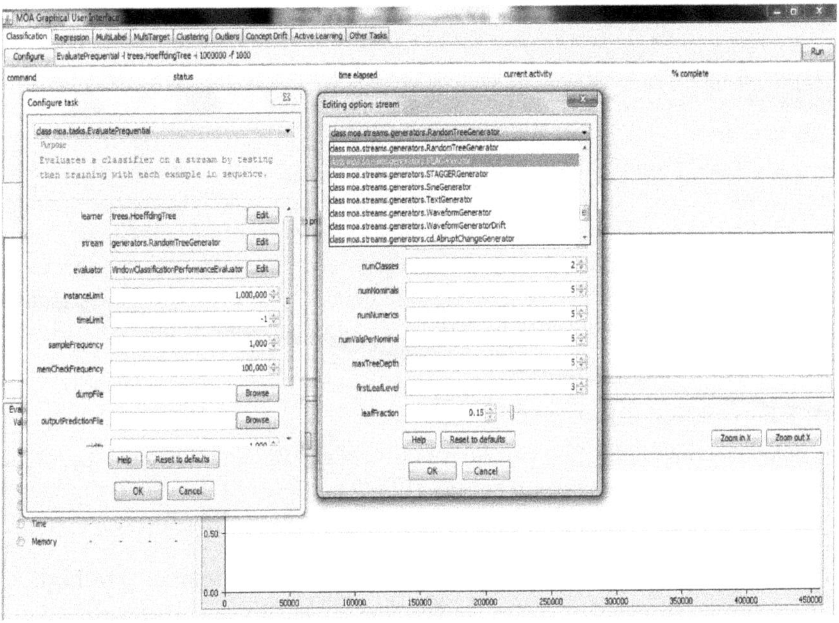

FIGURE 11.11 Stream generator algorithms in MOA.

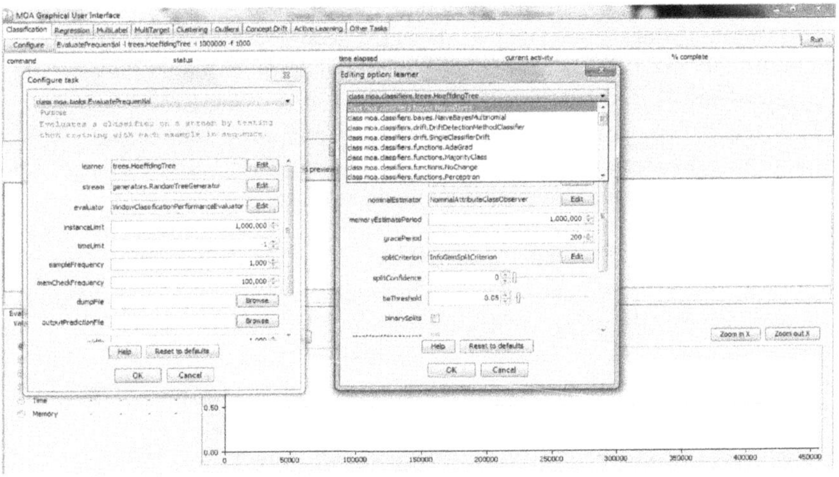

FIGURE 11.12 Classification algorithm in MOA.

technique has a list of inbuilt ML algorithms available for execution. This chapter mainly focuses on classification algorithms. Figure 11.12 shows various classification algorithms available in MOA. After clicking on 'Configure', the configure task window appears; when clicking on the 'Edit' button beside the learners, the Editing option learner window opens, and different classification algorithms appear.

11.5.4 Evaluation Methods

Classifier evaluation is one of the primary tasks in the ML process. The evaluation framework is composed of error estimation evaluation, performance metrics, statistically significant validation, and the cost measure of the process. An error estimation evaluation is a procedure that estimates the training and testing data. In batch learning, the dataset is split into test and train datasets. When the data in the dataset is limited, it recommends cross-validation. Stream learning involves a massive amount of unlimited data that arrives at a high speed. Stream learning poses different challenges. Cross-validation in streaming data is computationally too expensive and is not necessary. Two types of estimation approach arise, namely the Holdout Estimation Method and the Prequential Estimation Method.

11.5.4.1 Holdout Estimation Method

The holdout estimation method consists of predefined test and train sets that predict the current instance's accuracy precisely [12]. Obtaining test data from the streams is very difficult.

11.5.4.2 Prequential or Interleaved Test-Then-Train Estimation Method

In the Interleaved Test-Then-Train method, the model is tested with each instance before training. The accuracy of the model gets updated incrementally [12, 13]. The main advantage of using the interleaved test-then-train model is that testing does not need the holdout set. It uses the existing data for the analysis [12, 14]. In interleaved test-then-train, the available data calculates the accuracy. Still, in the holdout estimation method, the data available in the newest window is taken for the estimate of accuracy. Figure 11.13 shows the error estimation methods in MOA.

To access the error estimation method, click 'Configure', when the task window will appear. Now, click on the top drop-down menu, when various holdout, interleaved test-then-train error estimation methods will appear.

11.5.4.3 Evaluation Performance Metrics

In the case of data streams, the number of instances in each class keeps changing. At some point in time, the accuracy measures are appropriate only when there is a balanced number of instances in each class. The kappa statistic is another measure that deals with quantifying the predictive performance [12]. In addition to accuracy and the kappa statistic, other performance evaluation metrics are precision, recall, and F1 score values, and each of these performance metric values can be obtained from MOA. The comparisons are made on a given dataset, using various classification methods. Figure 11.14 shows the evaluation performance metrics in MOA.

Accuracy is the total number of correctly classified data relative to the total number of data points. The kappa statistic compares the observed accuracy with the expected accuracy, whereas precision shows how accurate the model is relative to the predicted positive instances. Recall calculates the number of accurately classified labeled positive instances. In the MOA framework, when 'Configure' is clicked, it opens an editing options evaluator window, which checks the performance metrics like precision, recall, and F1 score.

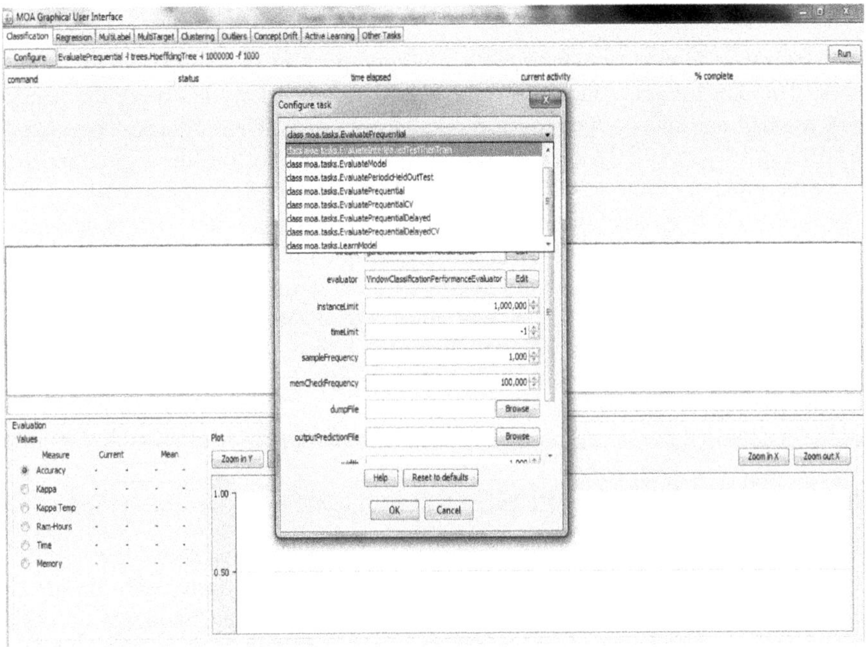

FIGURE 11.13 Error estimation methods in MOA.

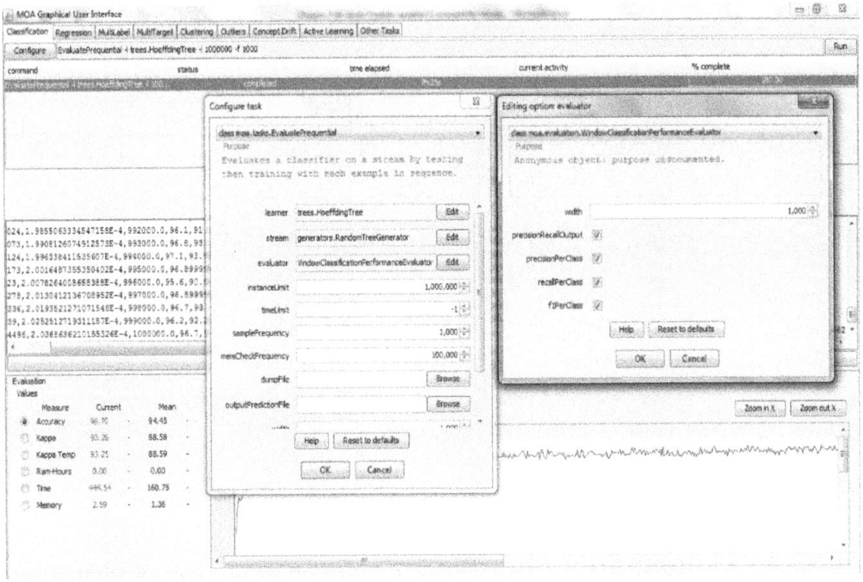

FIGURE 11.14 Evaluation performance metrics in MOA.

11.5.5 Discussions on Results

The dataset is loaded or generates a synthetic dataset, using the MOA framework. The dataset gets loaded into the MOA framework by applying the necessary algorithms and comparing the results obtained. The accuracy, precision, recall, and F1 scores of each algorithm are compared, to find the algorithm that gives the most accurate results. One cannot obtain the precision, recall, and F1 scores directly from the GUI window. The text document with these values must be saved to make these comparisons. The text document is opened in Excel format text. When the import wizard appears, select delimited and click 'Ok', then select tab and comma delimiters, and click 'Ok'. Finally, select general in column data format and click 'Finish'. Now, the results are loaded. Figure 11.15 shows the export text file tab, whereas Figure 11.16 shows how the results are stored in Excel.

11.6 LABORATORY EXERCISE AND SOLUTIONS

The goal is to compare the accuracies of the Hoeffding Tree and Naïve Bayes classifiers for a suicide dataset from Kaggle [15], with a file name master .arff file stream of 27,820 instances, using prequential evaluations. A sampling frequency of 1,000 instances is used to output the evaluation results every 1,000 instances.

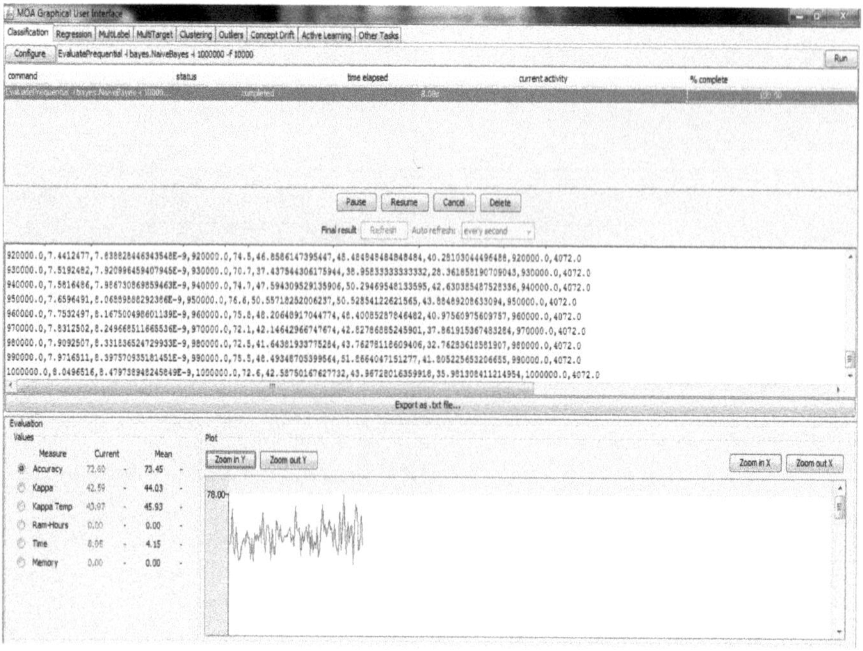

FIGURE 11.15 Export the text file.

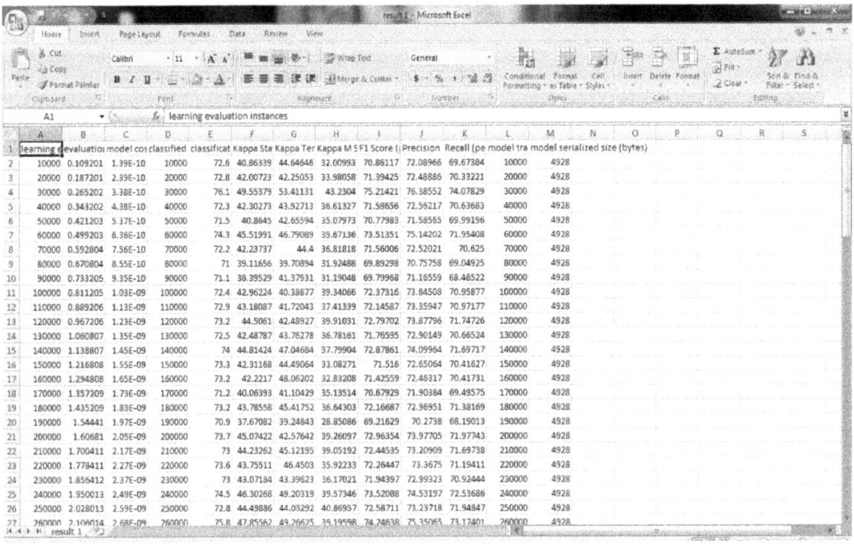

FIGURE 11.16 Exported text file result stored in Excel.

1. Load the dataset and run with Naïve Bayes Classifier. Figure 11.17 shows how to load the dataset and to run the Naïve Bayes classifier algorithm on the diabetes dataset.
2. Save the text file generated by using the Naïve Bayes algorithm. Rerun the dataset with the Hoeffding Tree algorithm. Figure 11.18 shows how the results are obtained by using the Hoeffding Tree algorithm.
3. Save the text file generated by the Hoeffding Tree. The MOA framework shows the comparison of the accuracy and other performance metrics of the Naïve Bayes and Hoeffding Tree algorithms. Figure 11.19 shows the comparisons of the two different classifiers, in which the Hoeffding Tree achieves greater accuracy than the Naïve Bayes algorithm.
4. Apart from the accuracy of the Naïve Bayes and Hoeffding Tree, the precision, recall, and F1 score are calculated. Figure 11.20 shows the precision, recall, and F1 score generated by the Naïve Bayes algorithm.
5. Precision, recall and F1 scores of Hoeffding Tree arc calculated. Figure 11.21 shows the performance metrics generated by the Hoeffding Tree algorithm.
6. Now, compare all the performance values from the Naïve Bayes and Hoeffding Tree algorithms. It evident that the Hoeffding Tree algorithm outperforms the Naïve Bayes algorithm (Table 11.2).

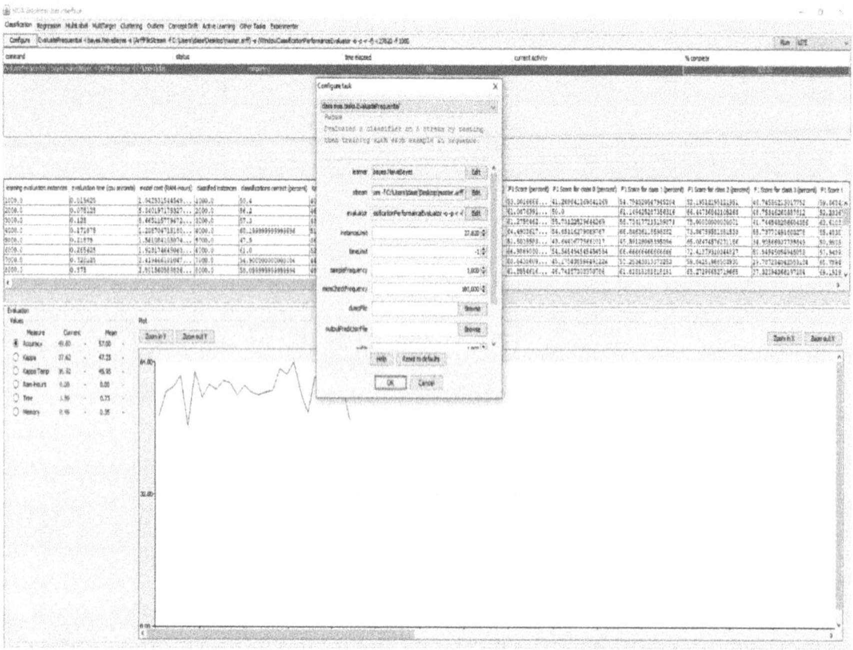

FIGURE 11.17 Loading the dataset into the configuration task window.

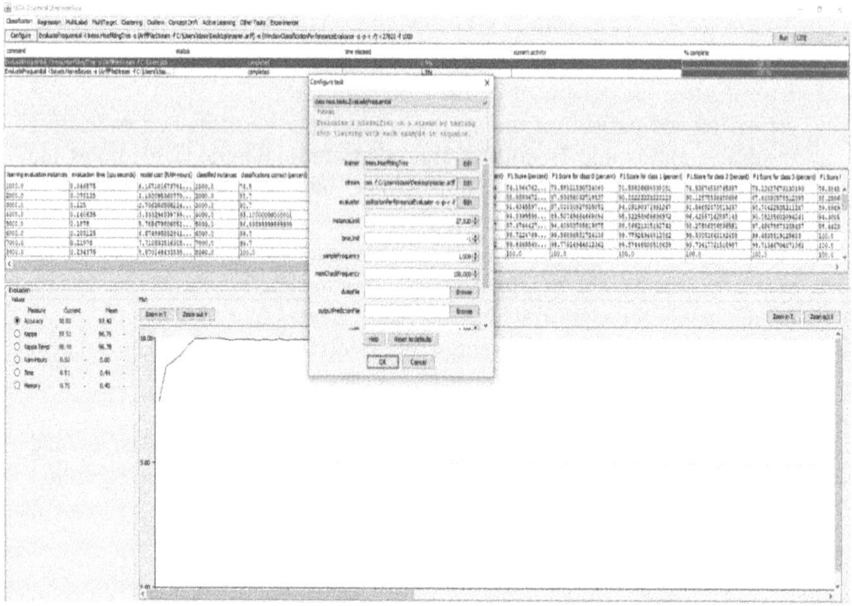

FIGURE 11.18 Result of the Hoeffding Tree algorithm.

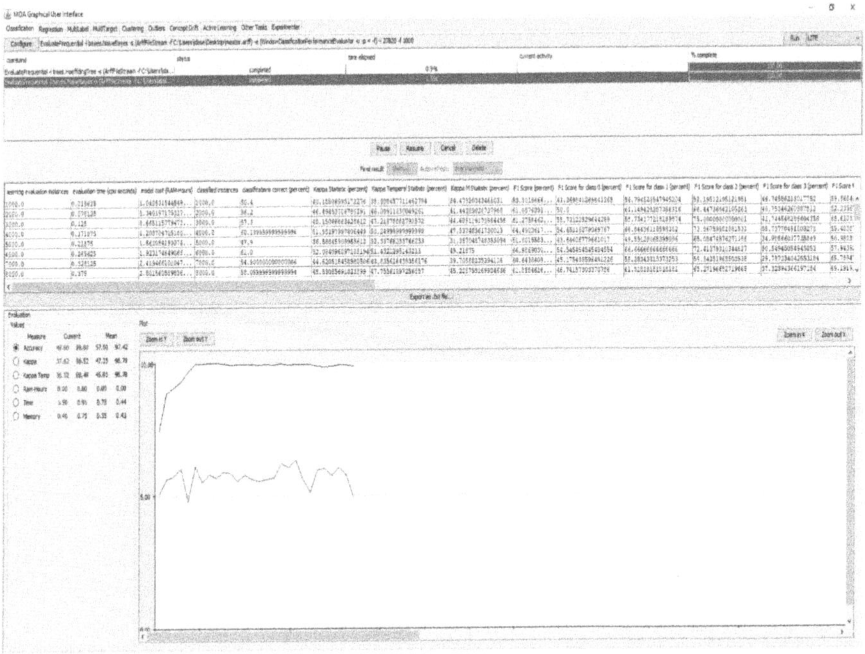

FIGURE 11.19 Comparisons of results from two different classifiers.

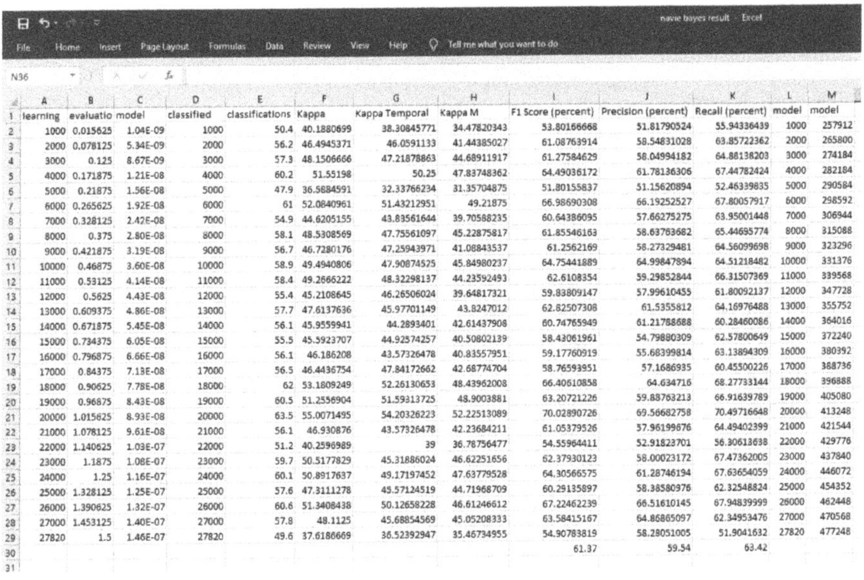

learning	evaluatio	model	classified	classifications	Kappa	Kappa Temporal	Kappa M	F1 Score (percent)	Precision (percent)	Recall (percent)	model	model
1000	0.015625	1.04E-09	1000	50.4	40.1880699	38.30845771	34.47820343	53.80166668	51.81790524	55.94336439	1000	257912
2000	0.078125	5.34E-09	2000	56.2	46.4945371	46.0591133	41.44385027	61.08763914	58.54831028	63.85722362	2000	265800
3000	0.125	8.67E-09	3000	57.3	48.1506666	47.21878863	44.68911917	61.27584629	58.04994182	64.88138203	3000	274184
4000	0.171875	1.21E-08	4000	60.2	51.55198	50.25	47.83748362	64.49036172	61.78136306	67.44782424	4000	282184
5000	0.21875	1.56E-08	5000	47.9	36.5884591	32.33766234	31.35704875	51.80155837	51.15620894	52.46339835	5000	290584
6000	0.265625	1.92E-08	6000	61	52.0840961	51.43212951	49.21875	66.98690308	66.19252527	67.80057917	6000	298592
7000	0.328125	2.42E-08	7000	54.9	44.6205155	43.83561644	39.70588235	60.64386095	57.66275275	63.95001448	7000	306944
8000	0.375	2.80E-08	8000	58.1	48.5308569	47.75561097	45.22875817	61.85546163	58.63763682	65.44695774	8000	315088
9000	0.421875	3.19E-08	9000	56.7	46.7280176	47.25943971	41.08843537	61.2562169	58.27329481	64.56099698	9000	323296
10000	0.46875	3.60E-08	10000	58.9	49.4940806	47.90874525	45.84980237	64.75441889	64.99847894	64.51218482	10000	331376
11000	0.53125	4.14E-08	11000	58.4	49.2666222	48.32298137	44.23592493	62.6108354	59.29852844	66.31507369	11000	339568
12000	0.5625	4.43E-08	12000	55.4	45.2108645	46.26506024	39.64817321	59.83809147	57.99610455	61.80092137	12000	347728
13000	0.609375	4.86E-08	13000	57.7	47.6137636	45.97701149	43.8247012	62.82507308	61.5355812	64.16976488	13000	355752
14000	0.671875	5.45E-08	14000	56.1	45.9559941	44.2893401	42.61437908	60.74765949	61.21788688	60.28460086	14000	364016
15000	0.734375	6.05E-08	15000	55.5	45.5923707	44.92574257	40.50802139	58.43061961	54.79880309	62.57800649	15000	372240
16000	0.796875	6.66E-08	16000	56.1	46.186208	43.57326478	40.83557951	59.17760919	55.68399814	63.13894309	16000	380392
17000	0.84375	7.13E-08	17000	56.5	46.4436754	47.84172662	42.68774704	58.76593951	57.1686935	60.45500226	17000	388736
18000	0.90625	7.78E-08	18000	62	53.1809249	52.26130653	48.43962008	66.40610858	64.634716	68.27733144	18000	396888
19000	0.96875	8.43E-08	19000	60.5	51.2556904	51.59313725	48.9003881	63.20721226	59.88763213	66.91639789	19000	405080
20000	1.015625	8.93E-08	20000	63.5	55.0071495	54.20326223	52.22513089	70.02890726	69.56682758	70.49716648	20000	413248
21000	1.078125	9.61E-08	21000	56.1	46.930876	43.57326478	42.23684211	61.05379526	57.96199676	64.49402399	21000	421544
22000	1.140625	1.03E-07	22000	51.2	40.2596989	39	36.78756477	54.55964411	52.91823701	56.30613638	22000	429776
23000	1.1875	1.08E-07	23000	59.7	50.5177829	45.31886024	46.62251656	62.37930123	58.00023172	67.47362005	23000	437840
24000	1.25	1.16E-07	24000	60.1	50.8917637	49.17197452	47.63779528	64.30566575	61.28746194	67.63654059	24000	446072
25000	1.328125	1.25E-07	25000	57.6	47.3111278	45.57124519	44.71968709	60.29135897	58.38580976	62.32548824	25000	454352
26000	1.390625	1.32E-07	26000	60.6	51.3408438	50.12658228	46.61246612	67.22462239	66.51610145	67.94839999	26000	462448
27000	1.453125	1.40E-07	27000	57.8	48.1125	45.68854569	45.05208333	63.58415167	64.86865097	62.34953476	27000	470568
27820	1.5	1.46E-07	27820	49.6	37.6186669	36.52392947	35.46734955	54.90783819	58.28051005	51.9041632	27820	477248
								61.37	59.54	63.42		

FIGURE 11.20 Precision, recall, and F1 score generated by the Naïve Bayes algorithm.

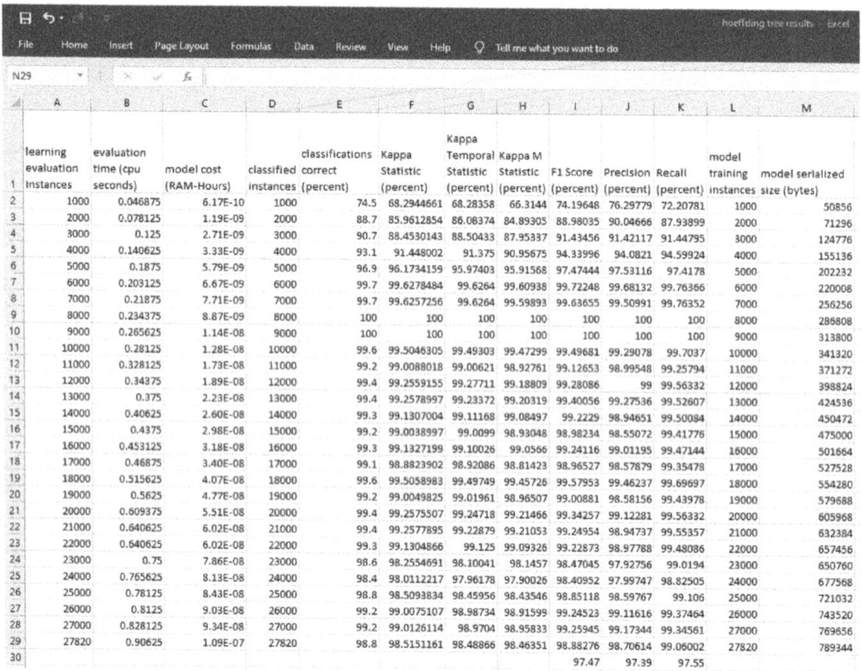

	learning evaluation instances	evaluation time (cpu seconds)	model cost (RAM-Hours)	classified instances	classifications correct (percent)	Kappa Statistic (percent)	Kappa Temporal Statistic (percent)	Kappa M Statistic (percent)	F1 Score (percent)	Precision (percent)	Recall (percent)	model training instances	model serialized size (bytes)
2	1000	0.046875	6.17E-10	1000	74.5	68.2944661	68.28358	66.3144	74.19648	76.29779	72.20781	1000	50856
3	2000	0.078125	1.19E-09	2000	88.7	85.9612854	86.08374	84.89305	88.98035	90.04666	87.93899	2000	71296
4	3000	0.125	2.71E-09	3000	90.7	88.4530143	88.50433	87.95337	91.43456	91.42117	91.44795	3000	124776
5	4000	0.140625	3.33E-09	4000	93.1	91.448002	91.375	90.95675	94.33996	94.0821	94.59924	4000	155136
6	5000	0.1875	5.79E-09	5000	96.9	96.1734159	95.97403	95.91568	97.47444	97.53116	97.4178	5000	202232
7	6000	0.203125	6.67E-09	6000	99.7	99.6278484	99.6264	99.60938	99.72248	99.68132	99.76366	6000	220008
8	7000	0.21875	7.71E-09	7000	99.7	99.6257256	99.6264	99.59893	99.63655	99.50991	99.76352	7000	256256
9	8000	0.234375	8.87E-09	8000	100	100	100	100	100	100	100	8000	286808
10	9000	0.265625	1.14E-08	9000	100	100	100	100	100	100	100	9000	313800
11	10000	0.28125	1.28E-08	10000	99.6	99.5046305	99.49303	99.47299	99.49681	99.29078	99.7037	10000	341320
12	11000	0.328125	1.73E-08	11000	99.2	99.0088018	99.00621	98.92761	99.12653	98.99548	99.25794	11000	371272
13	12000	0.34375	1.89E-08	12000	99.4	99.2559155	99.27711	99.18809	99.28086	99	99.56332	12000	398824
14	13000	0.375	2.23E-08	13000	99.4	99.2578997	99.23372	99.20319	99.40056	99.27536	99.52607	13000	424536
15	14000	0.40625	2.60E-08	14000	99.3	99.1307004	99.11168	99.08497	99.2229	98.94651	99.50084	14000	450472
16	15000	0.4375	2.98E-08	15000	99.2	99.0038997	99.0099	98.93048	98.98234	98.55072	99.41776	15000	475000
17	16000	0.453125	3.18E-08	16000	99.3	99.1327199	99.10026	99.0566	99.24116	99.01195	99.47144	16000	501664
18	17000	0.46875	3.40E-08	17000	99.1	98.8823902	98.92086	98.81423	98.96527	98.57879	99.35478	17000	527528
19	18000	0.515625	4.07E-08	18000	99.6	99.5058983	99.49749	99.45726	99.57953	99.46237	99.69697	18000	554280
20	19000	0.5625	4.77E-08	19000	99.2	99.0049825	99.01961	98.96507	99.00881	98.58156	99.43978	19000	579688
21	20000	0.609375	5.51E-08	20000	99.4	99.2575507	99.24718	99.21466	99.34257	99.12281	99.56332	20000	605968
22	21000	0.640625	6.02E-08	21000	99.4	99.2577895	99.22879	99.21053	99.24954	98.94737	99.55357	21000	632384
23	22000	0.640625	6.02E-08	22000	99.3	99.1304866	99.125	99.09326	99.22873	98.97788	99.48086	22000	657456
24	23000	0.75	7.86E-08	23000	98.6	98.2554691	98.10041	98.1457	98.47045	97.92756	99.0194	23000	650760
25	24000	0.765625	8.13E-08	24000	98.4	98.0112217	97.96178	97.90026	98.40952	97.99747	98.82505	24000	677568
26	25000	0.78125	8.43E-08	25000	98.8	98.5093834	98.45956	98.43546	98.85118	98.59767	99.106	25000	721032
27	26000	0.8125	9.03E-08	26000	99.2	99.0075107	98.98734	98.91599	99.24523	99.11616	99.37464	26000	743520
28	27000	0.828125	9.34E-08	27000	99.2	99.0126114	98.9704	98.95833	99.25945	99.17344	99.34561	27000	769656
29	27820	0.90625	1.09E-07	27820	98.8	98.5151161	98.48866	98.46351	98.88276	98.70614	99.06002	27820	789344
30									97.47	97.39	97.55		

FIGURE 11.21 Precision, recall, and F1 score generated by the Hoeffding Tree algorithm.

TABLE 11.2
Performance (%) Comparison

Performance Measures	Naïve Bayes	Hoeffding Tree
Accuracy	57	97.42
Precision	59.54	97.39
Recall	63.42	97.55
F1 Score	61.37	97.47

11.7 CONCLUSION

This chapter mainly focuses on the medical data generated from various devices, such as life support monitoring devices, wearable devices, wireless devices, etc. These data are further categorized as clinical data, omics data, or sensor data. All the data that are generated from the various devices are saved and used for the analyses. From the data obtained, the disease predictions are made, using ML techniques. This chapter mainly deals with the classification techniques in ML and tells how the classification techniques are used in the prediction of disease. This chapter explains in detail the Massive Online Analysis (MOA) framework. It shows how to load the dataset and to apply the classification algorithms. It explains how performance measures are obtained and compared. ML techniques are the foundation of the medical industry; to predict advanced diseases more accurately increases the quality and longevity of human life.

REFERENCES

1. J.G. Richens, C.M. Lee, S. Johri, "Improving the accuracy of medical diagnosis with causal machine learning," *Nature Communications*, vol. 11, pp. 1–9, 2020. doi: 10.1038/s41467-020-17419-7.
2. Dataflair TEAM·OCTOBER 19, 2019, Machine Learning in Healthcare – Unlocking the Full Potential! https://data-flair.training/blogs/machine-learning-in-healthcare/.
3. C. Réda, E. Kaufmann, A. Delahaye-Duriez,"Machine learning applications in drug development," *Computational and Structural Biotechnology Journal*, vol. 18, pp. 241–252, 2020. doi: 10.1016/j.csbj.2019.12.006.
4. S. Zeadally, F. Siddiqui, Z. Baig, A. Ibrahim, "Smart healthcare: Challenges and potential solutions using internet of things (IoT) and big data analytics," *PSU Research Review*, vol. 4, no. 2, pp. 149–168, 2019. doi: 10.1108/PRR-08-2019-0027.
5. Health Sciences Library, University of Washington, Dec 23, 2019, Data Resources in the Health Sciences https://guides.lib.uw.edu/hsl/data/findclin.
6. X. Ma, Z. Wang, S. Zhou, H. Wen, Y. Zhang, "Intelligent healthcare systems assisted by data analytics and mobile computing," *Wireless Communications and Mobile Computing*, vol. 2018, p. 16, July 2018.
7. A. Dhillon, A. Singh, "Machine learning in healthcare data analysis: A survey," *Journal of Biology and Today's World*, vol. 8, pp. 1–10, July 2019.
8. C. Bui, N. Pham, A. Vo, A. Tran, A. Nguyen, T. Le, "Time series forecasting for healthcare diagnosis and prognostics with the focus on cardiovascular diseases," In *6th International Conference on the Development of Biomedical Engineering in Vietnam (BME6), IFMBE Proceedings, Springer Nature Singapore Pvt Ltd*, vol. 63, p. 809, 2018.
9. Y. Tseng, X. Ping, J. Liang, P. Yang, G. Huang, F. Lai, "Multiple-time-series clinical data processing for classification with merging algorithm and statistical measures," *IEEE Journal of Biomedical and Health Informatics*, vol. 19, no. 3, pp. 1036–1043, May 2015.
10. P. Mishra, V. Varadharajan, U. Tupakula, E. S. Pilli, "A detailed investigation and analysis of using machine learning techniques for intrusion detection," *IEEE Communications Surveys & Tutorials*, vol. 21, no. 1, pp. 686–728, Firstquarter 2019. doi: 10.1109/COMST.2018.2847722.
11. Jason Brownlee on April 8, 2020, 4 Types of Classification Tasks in Machine Learning. https://machinelearningmastery.com/types-of-classification-in-machine-learning/.
12. A. Bifet, R. Gavalda, G. Holmes, B. Pfahringer. *Machine Learning for Data Streams with Practical Examples in MOA*. MIT Press, 2018. doi: 10.7551/mitpress/10654.001.0001
13. A. Bifet, G. Holmes, B. Pfahringer, P. Kranen, H. Kremer, T. Jansen, T. Seidl, "MOA: Massive online analysis, a framework for stream classification and clustering," *JMLR: Workshop and Conference Proceedings*, vol. 11, pp. 44–50, 2010.
14. P.K. Srimani, M.M. Patil, "Mining data streams with concept drift in massive online analysis frame work," *Wseas Transactions on Computers*, vol. 15, pp. 133–142, 2016.
15. Data Set from Kaggle. https://www.kaggle.com/russellyates88/suicide-rates-overview-1985-to-2016.
16. M. van Hartskamp, S. Consoli, W. Verhaegh, M. Petkovic, A. van de Stolpe, "Artificial intelligence in clinical health care applications: Viewpoint," *Interactive Journal of Medical Research*, 8(2), pp. 1–8, 2019.

12 Prediction of Coronavirus (COVID-19) Disease Health Monitoring with Clinical Support System and Its Objectives

G. S. Pradeep Ghantasala, Anu Radha Reddy, and M. Arvindhan

CONTENTS

12.1 INTRODUCTION

The impact of the COVID-19 pandemic infection [1] is increasing day by day, and people around the world are becoming more concerned. The practice of social distancing is to control the spread of the disease, as we discern the extent of the virus spread following specific rules based on various subjects, like mathematics, and information science for modeling the data. As we know, the behavior of individuals is not constant in nature, while the public is also adjusting to the current situation. Epidemiology is a discipline that deals with behavioral science, health services, research, and economics, together with medical investigation, which helps us to understand what the virus does to patients, to understand the genetic features of the disease, and to fight with the present pandemic disease.

There is the possibility of using a highly developed and sophisticated method of machine learning and data analytics [5] to merge all the different threads of the investigation. The current study is of relevance to policymakers around the world with respect to public health involvements, to help us get ahead of the disease development curve. Today, I will introduce you to these invaluable tools and focus on how to use them to tackle this pandemic situation, based on science. One area of interest concerns the neighborhood meeting, which is a crucial characteristic of earlier diseases like Ebola, swine flu, and SARS. This includes changes in the representation of the disease model. For example, let us consider how rapidly the disease is spreading and how it has been transmitted to reach the current situation. As we have in India a considerable number of research groups that specialize in the modeling of emerging infectious diseases, which are keenly supported and financed by governments, we plan to assist doctors and the public in this pandemic situation (Figure 12.1).

The purpose of writing about this problem is to discuss the transmission of the virus through the air as a transmission medium, as many researchers are performing different kinds of experiments on this disease by collecting data around the world. The sudden occurrence and spread of the new disease called the novel coronavirus (COVID-19), is the most remarkable communal health crisis of the 21st century [6]. People around the world are eagerly awaiting the development of treatments, such as vaccines, to control this disease. Some questions are related to science, such as how are such high numbers of people being infected? How dangerous is the virus? When will vaccines become available? How is the economy being impacted? What is the role of media channels, community media, and neighboring communities in the present outbreak situation? How long must lockdown be followed?

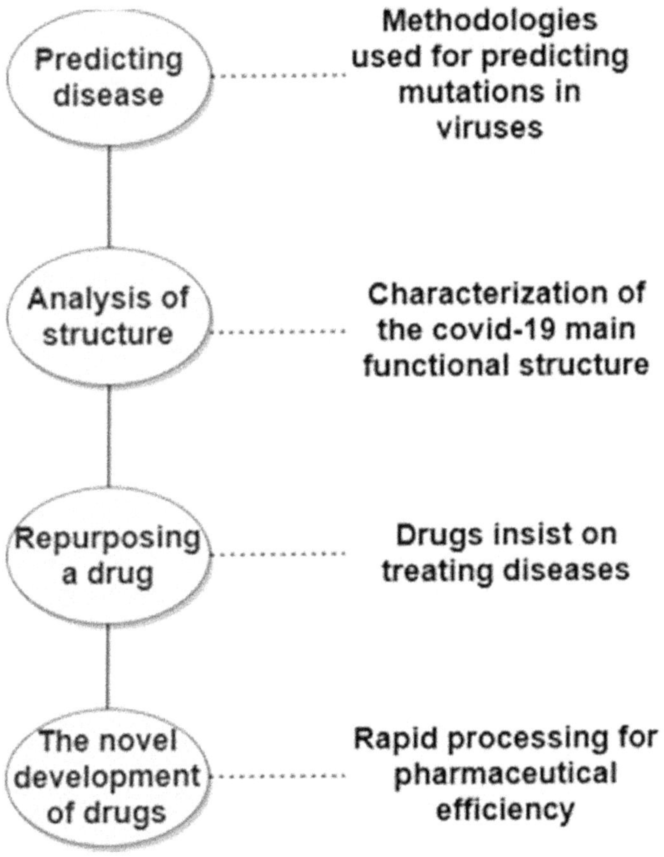

FIGURE 12.1 Emergence of the novel coronavirus.

The scope of this chapter is to provide an idea of using machine learning to help fight the pandemic situation. Samples of datasets are being collected from various countries and stored in a database. By using classification and linear regression techniques [5], we will show the statistics of the data collected from the samples in the form of a graph. Even though all the data on COVID-19 from across the world is being monitored and updated by the World Health Organization (WHO), here, the most important aspect we have to take into consideration is regular updating of situational reports into the database. All the regional health agencies across India are providing information relevant to their localities. The data are being collated into the massive databases that are helping to make records readily available for the researchers. The centers collating such data are Centers for Disease Control and Prevention (CDC) and Electronic Centres for Disease Control (ECDC) [7]. One of the best disease control research centers is that of Johns Hopkins University. It is located in Maryland, USA, where it maintains the live COVID-19 disease tracker, which provides regular updates on the rapidly changing situation in the progression

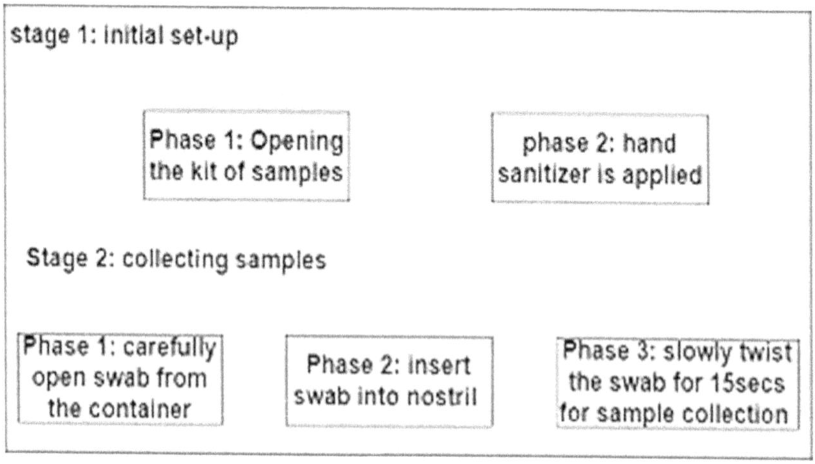

FIGURE 12.2 Screening a test sample for COVID-19.

of outbreaks around the world. Focusing primarily on collecting and organizing up-to-date information on how various countries are responding to the crisis from the aspect of the health system around the world, it is playing a crucial part in updating the virus data. One key question is what kind of data should be reported, and what should not be reported (Figure 12.2)?

12.2 HISTORY OF COVID-19

The COVID-19 pandemic, caused by the novel coronavirus, has led to massive changes, with scientific conference cancellations, social distancing, travel restrictions, lockdown, and many more unpredicted prevention measures. What is the reason behind this situation?

The sudden occurrence of the novel coronavirus was registered in Wuhan, Hubei Province, China, in December 2019 [8]. The pandemic has extended to all continents and countries in the world. The world's health systems are learning about the treatments against the virus and for the prevention of infections by the virus in humans, with epidemic data being released daily. In 2019, US President Donald Trump initiated an action plan, that was to be adopted under adverse circumstances, with the Ministry of Public Health and Human Services, to bring a successful conclusion to an epidemic over a wide geographical area, known as "Crimson Contagion" [9], which was supposed to be a viral pandemic starting in China and spreading around the world. The action plan proposed that 58,600 people might die across the United States [10].

A wide variety of restrictive measures have achieved high levels of control of COVID-19 in mainland China, South Korea, Hong Kong, and Singapore. In Europe, a comparatively huge population is experiencing harsh interventions, where people are going through multiple and prolonged lockdowns. If the gloomy model of the novel coronavirus disease comes true, then the model named "Crimson Contagion" will appear like a day in the park.

According to records on 26 March 2020, there were more than 450,000 confirmed cases of COVID-19 disease around the world, with more than 20,000 deaths affecting every continent, as this was a global pandemic before the World Health Organization announced it on 11 March 2020.

The novel coronavirus is an example of a well-known enemy, a pathogenic microorganism. Nothing has killed more human beings over historical time than the viruses, bacteria, and parasites that cause disease, although natural disasters, such as individual earthquakes, tsunamis, or volcanoes might cause more deaths (Figure 12.3).

Two critical factors must be considered to be essential to control this pandemic. The first factor is the provision of past information on coronavirus diseases to identify the origin of the virus and pandemic. The second factor involves the provision of instructions from the most reliable coronavirus references for the prevention and control of the virus in the workplace and at home (Figure 12.4).

12.2.1 Coronavirus

The coronavirus belongs to a large family of zoonotic viruses that causes illnesses ranging from the common cold to critical respiratory problems. Zoonotic viruses [11] can be defined as viruses which can be transmitted from animals to human beings. There are some familiar viruses spreading in distinct animal communities that have so far not been transmitted to humans. In the present scenario, COVID-19 is the most trending human disease around the world.

FIGURE 12.3 Doctor treating a COVID-19 patient.

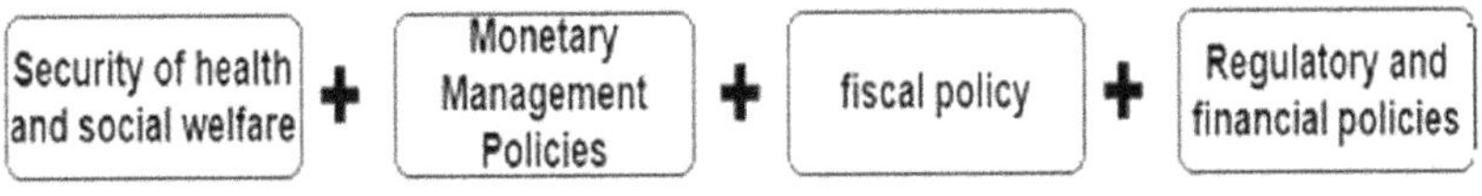

FIGURE 12.4 Economic restitution policy instruments following COVID-19.

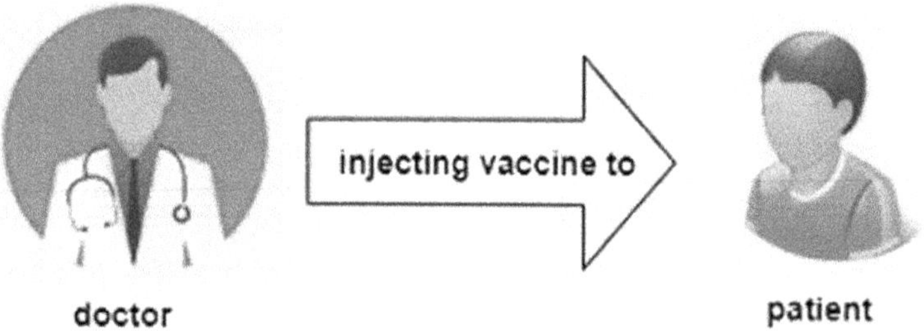

FIGURE 12.5 Doctor injecting patient with vaccine.

Common indicators of the COVID-19 pandemic infection are very close to those of the broadly distributed common cold and encompass respiratory symptoms, such as a dry cough, fever, headache, running nose, breathing difficulties, and shortness of breath. In most serious infections, the disease may cause pneumonia, acute respiratory problems, and side effects like kidney failure and the patient might die. The infection is viral, transmitted from one person to another *via* droplets produced frequently from the saliva of the respiratory system of the infected person during coughing and sneezing. According to the present data, the date from infection to the onset of symptoms is habitually between two and 14 days, with an average of five days. Shortly before the COVID-19 pandemic, two more coronavirus epidemics had occurred, namely MERS-CoV and SARS-CoV [12].

As we know, COVID-19 is affecting the entire global economy, which helps in the spread of new contagious diseases, and its long infective period of production is uniquely well suited causing a serious pandemic. Combined with the speed of modern human transportation, COVID-19 has the potential to reach practically any spot in the world, alive and transmissible, within 20 hours or less, contributing to its ability to spread, where similar viruses in the past would not have survived. For all the progress [13] we have made against infectious diseases, we are largely at the mercy of a microorganism that multiplies 40 million times more quickly than human beings do. Approximations of the economic cost of COVID-19 have so far passed the trillion-dollar mark (Figure 12.5).

Advances in vaccine skepticism has led the WHO in 2019 to call the anti-vaccination movement one of the planet's top 10 health threats [14].

12.2.2 Global Health Security

An international team of experts carried out a comprehensive evaluation of infectious diseases, including a survey of health reliability and response competence in 195 countries. The main aim of this project was to provide information from contagious disease epidemics, which could guide approaches to controlling international outbreaks [15] and estimate response capabilities for every state, in India. The aim was that the global health index would be foremost in making absolute changes in national health security and would improve the international state of being prepared for facing a particular epidemic situation.

The Global Health Index has guided various levels of response:

1. **Prevention:** The action of stopping a disease from becoming an epidemic.
2. **Observation and Announcement:** Preliminary observation and reporting of the widespread occurrence of an infectious disease of likely global threat.
3. **Quick Reaction:** Action of reducing the severity and seriousness of the disease outbreak.
4. **Fitness Method:** Development of an adequate and robust health system to serve the sick and safeguard the public.
5. **Approximation to Continental Standard:** Responsibility towards improving national income and financial plans, to show the way to close gaps and to keep operating to worldwide standards.
6. **Environmental Hazard:** Taking into account all the factors with respect to the environmental hazard, to provide a biological warning (Figure 12.6).

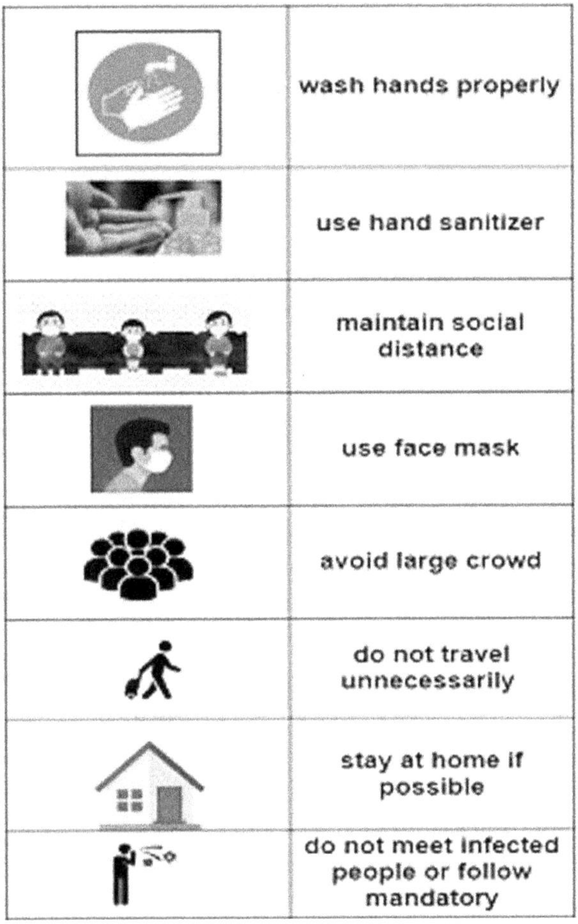
wash hands properly
use hand sanitizer
maintain social distance
use face mask
avoid large crowd
do not travel unnecessarily
stay at home if possible
do not meet infected people or follow mandatory

FIGURE 12.6 Safety measures against COVID-19.

12.2.3 Types of Coronavirus in Human Beings

In human beings, the virus is broadly classified into seven types, causing contagious viral infections. Symptoms resemble those associated with the common cold or influenza and may include:

- Nasal blockage
- Sore throat
- Cough
- Headache
- Fever

According to the US CDC, the COVID-19 virus causes problems by being carried in the mouth, nose, throat, and lungs, through which air acts as the transmission medium during breathing, as with pneumonia.

The symptoms of COVID-19aremost serious among:

- Infants
- Old people
- People with a weak immune system

All seven types of coronavirus that affect human beings are classified into two types (SARS and MARS).

12.2.3.1 General Coronavirus in Human Beings

According to the records of the WHO, coronavirus in humans is of four types:

- 229E
- NL63
- OC43
- HKU1

In general, coronavirus usually causes mild to severe health problems in humans.

All human beings in the world will spread at least one of the four previously mentioned viral infections during their life span. People who encounter this kind of virus are generally able to recover on their own.

12.2.3.2 Other Human Coronaviruses

Apart from the types of coronaviruses mentioned above, there are three different kinds of coronavirus (SARS-CoV, MERS-CoV, SARS-CoV-2) that were initiated as animal infections. These viruses develop over a long period and are, in the end, transmitted [16] to humans. Coronaviruses may pose a severe problem to human health. The animals most commonly affected by the coronaviruses are:

- Birds
- Bats

- Camels
- Pigs

12.2.3.3 SARS-CoV (Severe Acute Respiratory Syndrome Coronavirus)

According to the WHO, the first human case of SARS-CoV was registered in southern China in November 2002. SARS-CoV originated in bats and was transmitted to other animals before being transmitted to humans. During 2002–2003, approximately 8,000 humans in 26 countries across the world contracted SARS. According to health records, the total number of deaths registered was 774. The epidemic continued until mid-2003, with the execution of infection-reduction plans, such as isolation and quarantine, being effective. At present, there are no registered cases of SARS-CoV being transmitted anywhere in the world (Figure 12.7).

12.2.3.4 MERS-CoV (Middle East Respiratory Syndrome Coronavirus)

According to the WHO, the very first case of this virus in humans was registered in Riyadh, Saudi Arabia in September 2012; however, an unrelated case was recorded in March/April 2012 in Zarqa, Jordan. This virus was contracted by humans after exposure to camels which carried the infection. The transmission (camel–human, human–human) of the virus is also achieved by coming into very close contact with the breath of the infected person or animal. In 2012, 27 countries reported 2500 MERS cases, with the majority of the cases being filed in Saudi Arabia. In 2015, in South Korea, an outbreak of the disease registered 186 cases and 36 deaths. According to the CDC, the epidemic began with a person coming from China. According to the ECDPC, more than 300 cases of MERS-CoV were registered in 2019. The Ministry of Health Services is monitoring the pandemic situation around the globe.

12.2.3.5 SARS-CoV-2 (Severe Acute Respiratory Syndrome Coronavirus 2)

This virus is the cause of the novel coronavirus, COVID-19, which was first reported in Wuhan City, China, late in December 2019; later, officials of the Ministry of

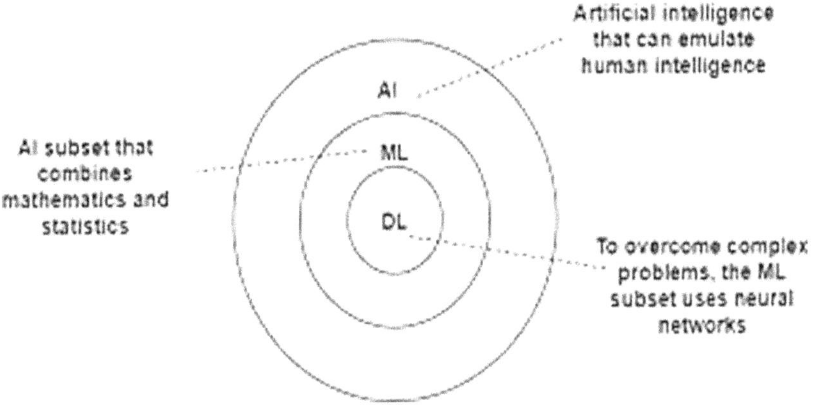

FIGURE 12.7 Relationship between artificial intelligence, machine learning, and deep learning.

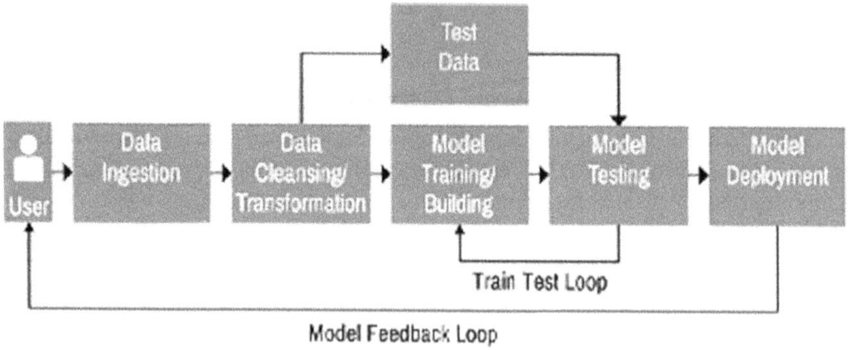

FIGURE 12.8 Model workflow of system.

Health in China observed a rise in the number of cases without knowing the exact reason for it, as the occurrence of the disease had been associated with a wet market selling fish, prawns, crabs, and chicken. However, the epidemic was likely started from an animal source, which has still to be identified. The virus spread across the world within a short period, spreading from infected person to healthy person contact through the medium of air. This virus can travel up to 8m into the air. The life span of this virus is 6 hours [17]. The following preventive actions were introduced globally to restrict the spread of COVID-19:

- Staying at home
- Washing hands frequently with soap for at least 20 seconds
- Using alcohol-based hand sanitizer
- Avoiding contact between hands and the face
- Maintaining social distancing
- Following regular updates of the local and international spread of the disease

The Indian government launched a mobile application associated with the coronavirus known as "Aarogya Setu". This app provides information regarding COVID-19, describing the precautions and the contact details of the helpline centers. It provides outbreak alerts and tracks live updates of COVID-19 (Figure 12.8).

12.3 INTER-RELATIONS BETWEEN ARTIFICIAL INTELLIGENCE, MACHINE LEARNING, AND DEEP LEARNING

Priyadharshini and Tagliaferri [37] categorize machine learning (ML) as a sub-discipline of artificial intelligence (AI), that permits a computer to train itself, as a result of exposure to a training dataset, without being programmed explicitly for this purpose. John McCarthy defined the term "Artificial Intelligence" in 1956 for a new field of study in the computer science domain, which focused on the creation of machines that imitated human cognitive function. An editorial by Puget (2016) reported that

Arthur Samuel coined the term "Machine Learning" in 1959, defining it as "a field of learning that gives the computer the capability to be trained without being explicitly programmed." According to a statement by Samuel (1959), it is feasible to plan a digital workstation to be trained accurately in the same way that humans or animals can be trained from previous experience, which would ultimately reduce the need for a comprehensive training effort. In 2006, Geoffrey Hinton coined the term "Deep Learning", described by Brownlee (2016) as a sub-field of machine learning, which facilitates the development of neural-like networks to replicate human-like decision making, wherein Neural Networks or Artificial Neural Networks are among the types of algorithms tasked with the action and structure of the human brain.

12.3.1 Machine Learning

As pointed out by Brownlee (2015), machine learning facilitates computers to train themselves. Compared with conventional programming, which merely represents computerization, machine learning is characterized by automating the method of training.

Information and programming are regarded as key to computers to achieve a predictable output throughout traditional methods, whereas machine learning takes in information with which to generate its own program. As stated by Prof. Mustafa (2012) in one of his lectures on The Learning Problem, conventional programming is successful only where a few mathematical formulae are necessary for solving a problem. Machine learning breaks this convention, working, even if it is not possible to pin down a mathematical principle, by repetitively recognizing the pattern and manipulating its algorithms as and when the data change. Prof. Mustafa additionally proved this by citing an example by predicting how an observer would rate a picture, relative to a prototype. Still, it is not likely to develop a mathematical solution because the way an individual would rate an image is not consistent with respect to how they would rate other images or how other people would rate the same image.

In 2016, Malhotra a machine learning researcher at Tata Consultancy Services (TCS), confirmed that data accessibility and estimation power are two of the attractive features that have led to a wide-ranging use of machine learning. Based on these two key inputs, machine learning models can extract patterns from huge volumes of data, a process which cannot be achieved manually, because a human brain cannot retain all the data for long and also cannot carry out such computations continually for hours.

12.3.1.1 Problem Types Solved through Machine Learning

Chen (2017), the editor of The Eliza Effect, distinguished five types of problems that can be resolved by applying machine learning:

- **Classification:** This approach is used to identify the grouping to which an object belongs. For example, if it is e-mail, is it spam? Or, if it is a tumor, is it cancerous?

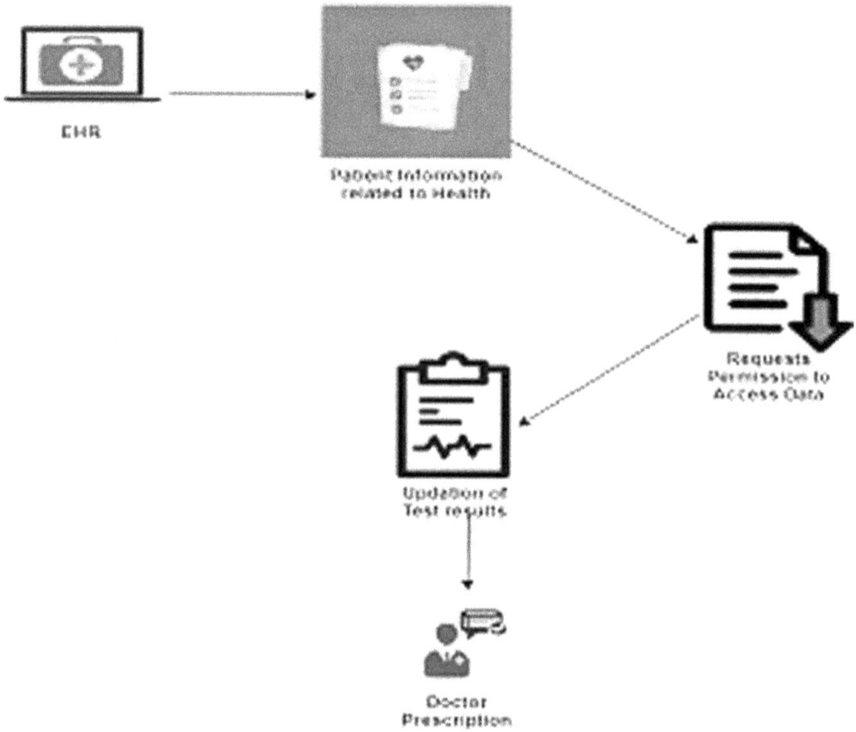

FIGURE 12.9 EHR reports.

- **Regression:** This approach is used to predict the value of a continuous numeric-valued variable related to another variable, such as the probability that a user would click on an advertisement or a stock price prediction.
- **Similarity/Anomaly:** This strategy is used to determine whether an instance belongs to the normal or standard grouping or to an anomalous grouping, such as when searching for similar images or to detect anomalies in user performance.
- **Ranking:** This approach is used to classify essential data according to a fuzzy input, such as Google Page grade.
- **Sequence Prediction:** This strategy is used to forecast the next element in a sequence of data, such as predicting the next word in a sentence (Figure 12.9).

12.3.1.2 Types of Machine Learning Algorithms

The three main categories of machine learning algorithms are supervised, unsupervised, and reinforcement learning.

12.3.1.2.1 *Supervised Learning*

The system is provided with label facts for both input as well as for predictable output during its training and the supervised learning algorithm generates a mapping function that can recognize the anticipated output for a piece of the given information. The training procedure continues until the algorithm reaches the chosen level of accuracy. One of the consistent goals of supervised learning is to create the computer knowledge of a categorization system; therefore, it is frequently used to solve classification problems. For example, the machine could be trained to classify a spam e-mail, distinguishing it from a valid e-mail, a strategy already being used by Google for Gmail spam filtering. *K*-Nearest Neighbor, Naïve Bayes, Decision Tree, Support Vector Machine, Linear Regression, and Neural Networks are a few of the simplest algorithms that are integrated into supervised learning.

12.3.1.2.2 *Unsupervised Learning*

The system is provided with an unlabeled and unclassified input dataset, and the unsupervised learning algorithm creates a reason to recognize concealed structures in the given dataset, based on patterns, similarities, and differences that exist among data, without any previous training. There is no evaluation of the level of accuracy of the structure or pattern recognized by the machine. Among the main focuses of unsupervised learning algorithms are clustering and association problems, and the most frequently used unsupervised learning algorithms include the *K*-Means algorithm for clustering and the Apriori algorithm for association problems.

12.3.1.2.3 *Reinforcement Learning*

The machine is open to the environment where it takes the decision on a test-and-fault basis and learns from its experience. For every correct decision, the machine receives a recompense response from the setting, that acts as a reinforcement signal, and the information about the rewarded state-action pair is stored. Later on, the mechanism iterates the rewarded behavior whenever faced with a similar situation. Reinforcement learning algorithms operate in a domain where tactical decision making is the input to success, like self-driving cars. Among the frequently used reinforcement learning algorithms are Q-Learning and Markov Decision Processes.

12.3.2 Machine Learning Workflow

Dimitrov (2017) depicted the machine learning workflow in the following description.

12.3.2.1 Smart Health Monitoring System

A wide range of health monitoring apps have been developed to supervise the health condition of patients. Health monitoring systems keep advancing and improving in this digital age [18]. Today, AI-powered, health-tracking, wearable devices endlessly monitor critical signs of a patient's health, with a considerable number of records per second. However, healthcare providers find it demanding to supervise several datasets across a large volume of patient data. To tackle this, machine learning apps in healthcare make available the identification of real-time activity in clinical systems, and further facilitate their interpretation and operation with earlier diagnosis and improved healing.

12.3.2.2 Electronic Health Records (Electronic Medical Records)

The purpose of electronic health records (EHR) is to provide a synopsis of how analysis of EHR data, using machine learning techniques, helps to monitor undiagnosed disease and to predict an emerging disease and its related complications. An EHR is a digital description of a patient's medical file. For it to be of value, the patient-centered medical record must be available immediately and accurately. An EHR consists of data from health check-ups and the treatment-related history of the patient to provide a broader vision of a patient's medical history. EHR can:

- Contain information related to the patient's previous health record, analysis, prescriptions, treatment plans, follow-up visits, allergies, and laboratory and investigation results.
- Contain the consent form for permission to access the database that the provider can use for decision making regarding a patient's condition.

Healthcare systems around the world use Electronic Medical Records (EMR), providing a longitudinal series of statistics about each patient's health. Exploitation of electronic records depends on the ease of access to and analysis of the enormous volume of data generated. In addition to the clinical value inherent in the EMR [19] and their potential for facilitating care and appropriate treatment for the patient [20], the sensitivity of the personal data contained in the EMR means that privacy of the individual must be protected. In particular, the uses of narrative deep learning methods are designed to hold multi-dimensional, heterogeneous records. These improvements are forecast to achieve an escalating uptake inaccuracy of medication during enlargement of modified health services facilitated by investigation of person as well as aggregate multimodal data reside keen on patient's EMR.

The most influential application of EMR information is to improve the treatment plan for each patient, because of the ability of machine learning to extract medical information from enormous datasets. An "accuracy medicine" proposal has been launched through a $255 million development to move medical performance away from a "one-size-fits-all" system [20] to more personalized medicine. Whereas EMR data are pooled with added modalities of information, such as molecular basis of facial appearance, biosensors, social determinants of health, and environmental exposure, among others, it will now be possible to characterize a patient's condition in many dimension and at numerous scales. Innovative strategies, such as the adapted cancer therapy agenda, have exploited EMR to improve treatment plans.

Similarly, added data parameters involving EMR are used to supplement the existing customary Modified Early Warning Score (MEWS) algorithms [21], based on the"track-and-trigger" warning, of patient status worsening based on six cardinal signs, in combination with permanent monitoring of patients. Additionally, developments from a data-driven, EMR-based study can lead to actionable results, such as

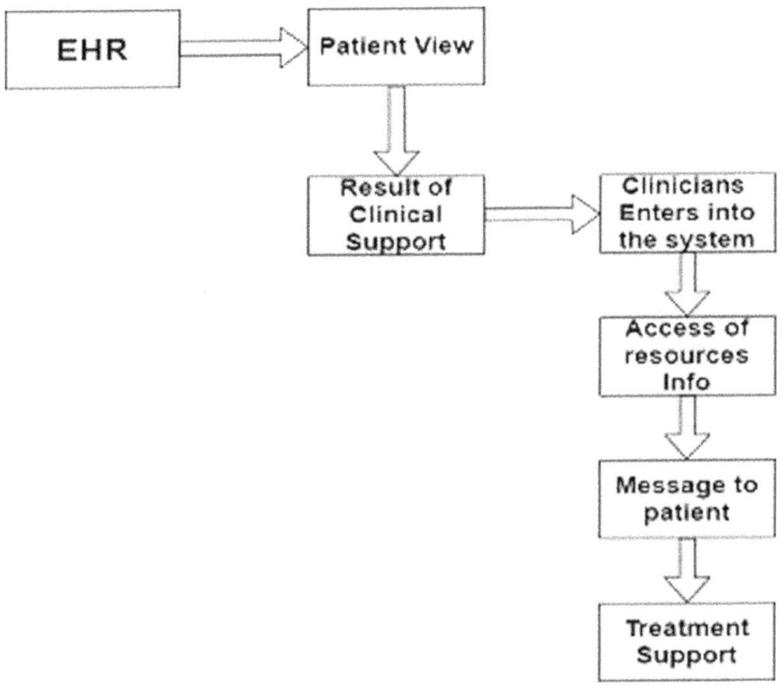

FIGURE 12.10 Clinical support system for patients.

identifying unpleasant prescription side effects and predicting future infection risks (Figure 12.10).

Furthermore, EMR is an authoritative tool to aid primary investigations [22]. There are many examples of research studies that identified patterns of syndrome receptiveness, co-morbidity, and infection trajectory based on EMR. Relating EMR to -omics record types contained in a system framework has enabled clinicians to identify diverse risk factors for syndrome etiology, including inherited, environmental, and demographic factors, and combinations there of. Hereditary and genomic information are frequently present in medical records, using hospital-affiliated BayBanks, to hire individuals from the patient population to acquire and "bank" their genetic information [23].

12.3.2.3 Manipulation of Supervised Concern and the Incorporated Delivery Scheme

The increased use of machine learning-based wearable devices and wireless monitors has established fewer notice from the business hawkers than the hospital testimony for the reason that of the difference in finance and rigid necessities. Information supervision tools, that make supervision of patients outside hospital surroundings easier and more effective, can assist the healthcare provider in monitoring a patient's condition [24, 25]. Reports from wearable devices may generate large amounts of data in addition to a varied set of additional data elements, such as X-ray images and pathology reports.

12.3.2.4 Functional Operation of an Electronic Health Record System

An EHR is not merely an electronic description of the hardcopy document. When the documentation covers measurement of a wide-ranging EHR system, it is also necessary for the electronic system to generate a straightforward message to facilitate decision making. A mechanism summarizing [26] a complete EHR system and demonstrating its functionality is necessary for full exploitation of the EHR. The five serviceableworkings are as follows:

- Integration of patient information
- Maintenance of medical data
- Systematic entry of clinicians' records
- Admittance to acquaintance possessions
- Use of the information to consolidate a treatment plan

12.3.2.4.1 Integration of Patient Information

An integrated summary of each patient's clinical and medical data is the most important reason for an EHR. Even though this might appear to be a comparatively trivial goal, attainment and accessibility of these records are significant problems to be overcome, such as the complications associated with the presentation and statistical analysis of the data [27], ranging from summary statistics to a graph or bar chart to descriptions of inter-variable relationships, and the sheer extent of the numerical data and records, such as those from medical laboratories, images from X-rays, magnetic resonance imaging (MRI), and computed tomography (CT), prescription records from public pharmacies, etc. Moreover, no exclusive national patient identifier classifier is present in the United States for associating patient information from numerous locations to the account of a single patient. The fact that diverse patient records use different unique identifiers for the same patient, data content terminologies, and data formats, creates extensive additional work.

Supervisors of each EHR scheme must amend the message format and the map code system from the basic plan to the order and regulations suitable for their EHR system. Nowadays, the majority of the medical data source can distribute the quantifiable content as health level communication. Still, the sender diverges from the typical and uses limited code as identifiers for notes and instructions in these letters. The record edge depicts not only the message-handling ability but also manually interprets laws from the foundation method to the ideal system of the receiving EHR.

The boundary locomotive makes available an exact and version shield among systems feigned by particular purveyors. In this approach, an organization can blend unusual vendors' goods and silently attain the objective of incorporated admission to unwearied records on behalf of the clinician. Clinicians need access to patient information in a flow sheet format to emphasize change over time with respect to several variables.

12.3.2.4.2 Maintenance of Clinical Data

It would be ideal to supply the most support to the patients with the most difficult and serious conditions, when the doctor is putting together her/his opinion of the patient's condition and is constructing a treatment plan. The mainly successful decision-support involvements comply with the recommended plan, while still allowing the doctor to oversee the final decision. As long as admittance to a short basis with the advice may enlarge receipt of reminders and at the similar time instruct the concerned contributor [28].

The patient analysis uses refined practices, depending on an extensive range of medical testing, to advise on, for example, antibiotic options, quantity given, and the duration of treatment. Clinicians can analyze the basis for the recommendation and the reason given. A prominent aspect of the procedure is its request [29] for a response when the clinicians decide not to follow the proposal. The advice is used to improve the chosen procedure and the associated software.

Online assessment on the selection of antimicrobials has resulted in markedly improved medical and economic outcomes for patients, as a result of improved management of communicable diseases. Reminders and alerts can be raised throughout the outpatient report. The system search for an appropriate decision-support policy and important print reminders on the outcome should be printed out ahead of the planned appointment.

12.3.2.4.3 Systematic Entry of Clinicians' Records

If the final objective of an EHR system is to assist clinicians in making an enlightened decision and appropriate treatment plan, then the scheme ought to produce the result as near as possible to the time of entry of the patient to the clinical system. Several methods have the potential to reach a conclusion throughout the order-entry progress [30]. The WIZ organize monitor put together in order regarding a patient's lively orders, a clinical alert based on existing information from the electronic patient record, and an abstract of relevant elements from the written notes.

A clinical alert, developing from the result of a laboratory analysis, can consist of a suggestion for suitable actions. Surgeon order-entry systems can warn the doctor with respect to potential problems with possible allergy [31] or drug response or inter-drug interactions ahead of producing the prescription summary, as shown on the monitor, regarding the patient's outpatient medical record. Formerly, a surgeon order-entry system is adopted into the, taking into account ethnicity, and/or altering the drug or dose in question, based on the most recent systematic assessment, can appreciably change the physician's treatment plan. Clinical excellence and treatment expenses can readily be changed.

12.3.2.4.4 Admittance to Acquaintance Resources

The nearly all query of information possessions, whether they are pleased by discuss with one more person assistant [32] or by penetrating through allusion resources or the copy, are conducted in the presence of a detailed patient record. Accordingly, the most efficient way to make available the right to use information is individual and is considered by the clinician. Nowadays a wide variety of available information sources range from the papers published in Pub Med to full-text assets such as COVID-19 [33], and online references such as advanced are obtainable for examination. As a result, it is moderately straightforward for the physician to find check-up data at the same time as appraising the outcome, script comments, or instructions online. Nonetheless, the lively appearance of text appropriate to an exacting clinical state, such as an "Info button", would strengthen the possibility that the data will change the physician's decision.

12.3.2.4.5 Use of Information to Consolidate a Treatment Plan

Seeing that the goal can become more and more dispersed in the midst of multidisciplinary healthcare professional, the efficiency and competence of messaging can influence the general coordination and the aptness of the care provided. Message tools ought to be incorporated within the EHR system, such that communications are automatically incorporated into a patient's documentation, i.e., the patient's record must be accessible as part of the visit [34].

Physical separation of panel member sites means that the information on the network must be accessible at every site in order to generate an appropriate decision on the patient's status. These sites consist of the providers' office, the sanatorium, the emergency room, and the patient's residence. Proximity to the patient's residence will enable a means of transportation for monitoring health and to enable usual contact with the patient. Communiqués also can be "hard-copied" to the patient through electronic messages, text or pager services, or "pulled" by a provider at their regular connections with the PC [35]. An EHR system can also be facilitated through regular patient handovers, to ensure the seamless transfer of information from one clinician to another. Classically, a concise oral or paper report will help the receiving clinician understand the patient's condition, as it is significant for building decisions and treatment plans while the more senior clinician is busy. While an in-person visit regularly defines a patient encounter, clinician decision making can occur in response to other contacts, such as a patient phone call concerning unusual symptoms, medicine replenishment needs, and the arrival of investigation results. Preferably, the clinician or a critical administrative center team should notify these actions and have the necessary EHR tools ready to react, together with the patient's electronic chart, a mechanism for automatic replenishment [36] authorizations, a template for producing information to the patient regarding test results, and a back-to-work form.

Furthermore, as the doctor asks the patient to attend an analytical check, such as a mammogram, an EHR system can maintain follow-up of the occasion and can report to the doctor that an investigation outcome has not been delivered by a particular time. These pathway functions [37] prevent analytical test results from"falling between the cracks". EHRs are usually stored by the organization in which they exist.

Communication apparatus [38], to maintain suitable and well-organized contact between the patient and the healthcare group, can improve overview and supervision of the patient's status. The EHealth [39] app can make their EHR available to patients and provide an online contact system by which the patient can readily ask a question about their check-up or ask about additional clinical (e.g., renewal of recommendation) or managerial tasks (e.g., book an appointment). In addition to sustaining contact between healthcare professionals and their patients, community-based EHRs can simplify the well-organized establishment and transmission of reports to maintain patient security, while improving public health and healthcare operation for individuals [40].

12.3.2.5 Inquiry and Inspection Systems

The capacity of computer-stored databases is not matched by physical systems. Medical staff, data security professionals, and administrators should be able to exploit this capability to examine patient outcomes and follow patterns. The public health professional is ready to use the treatment function of computer-stored databases for observation, to look for the emergence of a new disease, or for additional physical conditions that may require further medical consideration.

Even though these functions are dissimilar, their end goal is related. Together, the main aim is to inspect the patient's medical evidence and, if the evidence meet prescribed criteria, to create a correct diagnosis and develop an appropriate treatment plan. Queries usually deal with a big split, or every, of patient inhabitants; the production is a tabular report of chosen unprocessed facts on the entire the enduring reports repossess or a geometric synopsis of the value enclosed in the records. Supervision typically addresses merely a person's under lively heed; its output is a vigilant or prompt memo. This system can be used for irrefutable concerns, clinical study, demonstration study, with the admin.

12.3.2.6 Medical Care

Computerized reminders can improve the physicians' ability to look after their patients. A computerized system can recognize patients who are overdue for cyclic test examinations, such as immunization or cervical Pap (**Papanicolaou**) test, and can remind the physician to carry out these procedures at the next appointment. These reminder systems are particularly helpful to recognize and alert patients who need to receive a booster treatment drug. Such a reminder system method can identify patients who need to be subjected to follow-up examinations, or to collect the records necessary from patients to complete an audit.

12.3.2.7 Experimental Study

A computerized inquiry system is capable of identifying patients who meet certain eligibility criteria for a forthcoming medical trial. For instance, a researcher might recognize all patients in a check-up consulting room which exhibit the required characteristics whilst not having one or more exclusionary criteria. Observation can check the selection of trial patients during their visit, by following the steps of a "medical audition" to make sure that the treatment is specified.

An admission of soreness on the other side, for instance, might stimulate a survey concerning a continued investigation of a treatment for back ache. If the doctor concurs, the computer forwards an electronic file to the nurse recruiter, so that she/he can indicate to the health center whether the patient should be included into the revised trial.

12.4 CONCLUSION

The pandemic has had a serious effect on the lives of people around the world. A huge number of deaths has occurred, increasing every day. On the other hand, artificial intelligence and machine learning are proving to be highly effective in educational institutions and in our daily lives, with many successes. These techniques have also contributed to helping humans during this lockdown period while helping in the COVID-19 crisis response, e.g., by efficiently delivering educational course contents to students by on-line teaching through various cloud applications remotely from the universities. Artificial intelligence and machine learning play a vital role in medical diagnosis based, for example, on interpretation of radiology images of the chest, modeling of the transmission and spread of the virus, and making predictions based on the number of cases recorded in a particular period. The patient who is affected by the pandemic and admitted to hospital will receive good care and be treated according to a plan prescribed by the doctors. Health records are updated regularly and maintained to study the patients' health and mental status. Appointments for counseling at the scheduled time are arranged for the patient to recover from the mental aspect of the illness at an early stage. The doctors expend much effort to treat patients during this pandemic period. As we know, there no treatment available on the market to treat COVID-19, but the doctors are continuing to save our lives by putting their lives at risk to treat us, under conditions designed to minimize transmission. Finally, we must feel proud of the doctors, hospital staff, police department, and sanitary workers. They are working 24/7 during this epidemic situation, leaving their families and doing important services for our nation, without hesitation or exclusions.

REFERENCES

1. Agency for Healthcare Quality and Research (February 2014). Computerized Provider Order Entry. Health Information Technology. Retrieved December 10, 2014 American Medical Association. (2014). EHR Selection Considerations.
2. HealthInformation Technology, 8–11. Bell, Karen M., & Patterson, W. David. (2014). Understanding EHRs: Common Features and Strategic Approaches for Medicaid/ SCHIP.
3. Centers for Medicare & Medicaid Services. (2014, 02/26/2014 1:44 PM). E-Prescribing. Retrieved December 10, 2014. http://healthit.ahrq.gov/sites/default/files/docs/page/ understanding_ehrs___common_features_and_strateg ic_approaches_for_medicaid _schip_5.pdf
4. Donaldson, Molla S., Yordy, Karl D., Lohr, Kathleen N., & Vanselow, Neal A. (1996). Primary Care: America's Health in a New Era. Washington, D.C.: Institute of Medicine, National Academy Press. Gateway Community Health Center, Inc. (Producer). (2004, October 15, 2014). Gateway Community Health Center, Inc. Organizational Chart. http://www.cms.gov/Medicare/EHealth/Eprescribing/index.html?redirect=/ eprescribing
5. Halbesleben, Jonathon R. B., Wakefield, Douglas S., Ward, Marcia M., and Crandall, Donald (Jan 2009). Staff "super users" who train others on clinical information systems help shape positive employee attitudes. (09-RA010, No. 347), 82–96. http://www.gatewaychc.com/images/organizational chart2004.pdf

6. Health Resources and Services Administration. (2014a). Common EHR Functions. Retrieved December 10, 2014. http://archive.ahrq.gov/news/newsletters/researchactivities/jul09/0709RA26.html
7. Health Resources and Services Administration. (2014b). What is "Meaningful Use"? Retrieved December 10, 2014. http://www.hrsa.gov/healthit/toolbox/healthitimplementation/implementationtopics/selectcertifiedehr/sele ctacertifiedehr_5.html
8. HealthIT.gov. (2014a). CDS Starter Kit: Diabetes follow-up care. http://www.hrsa.gov/healthit/meaningfuluse/MUStage1CQM/whatis.html
9. HealthIT.gov. (2014b). Certification Programs & Policy. http://www.healthit.gov/sites/default/files/del-3-7-condition-example-diabetes.pdf
10. HealthIT.gov. (2014c). Certified EHR Technology and Certification Process. http://www.healthit.gov/policyresearchers-implementers/certification-programs-policy.
11. Primary Care Practice Facilitation Curriculum MODULE 26 M26–23HealthIT.gov. (2014e). Meaningful Use Definition and Meaningful Use Objectives of EHRs. Retrieved December 10, 2014. http://healthit.gov/policy-researchers-implementers/certified-health-it-product-list-chpl
12. HealthIT.gov. (2014f). Regional Extension Centers (RECs). http://www.healthit.gov/providers-professionals/meaningful-use-definitionobjectives
13. HealthIT.gov. (2013). What is an electronic health record (EHR)? Retrieved December 10, 2014, from http://www.healthit.gov/providers-professionals/faqs/what-electronic-health-record-ehr Hostetter, Martha. (2014). Case Study: Alabama's Together for Quality Program--Putting Health IT to Work for Medicaid Beneficiaries. http://www.healthit.gov/providersprofessionals/regional-extension-centers-recs.
14. Hsiao, C, Jr, &Hing, E. (2014). Use and characteristics of electronic health record systems among office based physician practices: United States, 2001–2013. Data Brief. Hyattsville, MD: National Center for Health Statistics. Institute of Medicine. (2001). http://www.commonwealthfund.org/publications/newsletters/quality-matters/2009/july-august-2009/casestudy
15. Knox, Lyndee, & Brach, Cindy. (2013). Crossing the Quality Chasm: A New Health System for the 21st Century. 1–8. Module 17. Electronic Health Records and Meaningful Use. Retrieved October 9, 2014.
16. Massachusetts eHealth Institute (2014). http://www.ahrq.gov/professionals/preventionchroniccare/improve/system/pfhandbook/mod17.html
17. Nemeth, Lynne S., Ornstein, Steven M., Jenkins, Ruth G., Wessell, Andrea M., & Nietert, Paul J. (2012). The Massachusetts eHealth Institute - Health Care Technology and Innovation. Retrieved December 10, 2014, from http://mehi.masstech.org/
18. Implementing and Evaluating Electronic Standing Orders in Primary Care Practice: A PPRNet Study. http://www.jabfm.org/content/25/5/594.long. doi:10.3122/jabfm.2012.05.110214. New York eHealth Collaborative. (2014). What We Do. Retrieved December 10, 2014, from http://nyehealth.org/what-we-do/ PCC EHR Solutions. (2014). Meaningful Use Measures Report.
19. http://learn.pcc.com/Content/PCCEHR/Reports/MUReport.htm Recovery.gov. (2014). The American Recovery and Reinvestment Act: The Recovery Act. Retrieved December 10, 2014, from http://www.recovery.gov/arra/About/Pages/The_Act.aspx.
20. T. Botsis, G. Hartvigsen, F. Chen, C. Weng. Secondary use of EHR: Data quality issues and informatics opportunities. Summit Transl. Bioinform. 2010 (2010), p. 1.
21. S.M. Meystre, G.K. Savova, K.C. Kipper-Schuler, J.F. Hurdle. Extracting information from textual documents in the electronic health record: A review of recent research.
22. M. Jiang, Y. Chen, M. Liu, S.T. Rosenbloom, S. Mani, J.C. Denny, H. Xu. A study of machine-learning-based approaches to extract clinical entities and their assertions from discharge summarie. *J. Am. Med. Inform. Assoc.*, 18 (5) (2011), pp. 601–606.

23. F. Doshi-Velez, Y. Ge, I. Kohane. Comorbidity clusters in autism spectrum disorders: an electronic health record time-series analysis. *Pediatrics*, 133 (1) (2014), pp. e54–e63. doi: 10.1542/peds.2013-0819.
24. S. Ebadollahi, J. Sun, D. Gotz, J. Hu, D. Sow, C. Neti. Predicting patient's trajectory of physiological data using temporal trends in similar patients: a system for near-term prognostics, In: *AMIAAnnualSymposiumProceedings*, Vol. 2010, American Medical Informatics Association, 2010, p. 192.
25. R. Miotto, C. Weng. Case-based reasoning using electronic health records efficiently identifies eligible patients for clinical trials. *J. Am. Med. Inform. Assoc.*, 22 (el) (2015), pp. e141–e150.
26. Hippisley-Cox, C. Coupland. Predicting risk of emergency admission to hospital using primary care data: derivation and validation of QAdmissions score. *BMJ Open*, 3, p. 8.
27. F. Rahimian, G. Salimi-Khorshidi, J. Tran, A. Payberah, J.R. Ayala Solares, F. Raimondi, M. Nazarzadeh, D. Canoy, K. Rahimi. Predicting the risk of emergency hospital admissions in the general population: development and validation of machine learning models in a cohort study using large-scale linked electronic health records. *PLOS Med.*, 15 (11).
28. T. Pham, T. Tran, D. Phung, S. Venkatesh. Predicting healthcare trajectories from medical records: a deep learning approach. *J. Biomed. Inform.*, 69 (2017), pp. 218–229. doi: 10.1016/j.jbi.2017.04.001
29. D. Ravì, C. Wong, F. Deligianni, M. Berthelot, J. Andreu-Perez, B. Lo, G. Yang. Machine learning for health informatics. *IEEE J. Biomed. Health Inform.*, 21 (1) (2017), pp. 4–21. doi: 10.1109/JBHI.2016.2636665.
30. E. Gawehn, J.A. Hiss, G. Schneider. Machine learning in drug discovery. *Mol. Inform.*, 35 (1) (2016), pp. 3–14.
31. E. Choi, M.T. Bahadori, A. Schuetz, W.F. Stewart, J. Sun. Doctor AI: Predicting Clinical Events via Recurrent Neural Networks, In *Proceedings of the 1st Machine Learning for Healthcare Conference, Vol. 56 of Proceedings of Machine Learning Research, PMLR*, 2016, pp. 301–318. http://proceedings.mlr.press/v56/Choi16.html.
32. B.C. Kwon, M.-J. Choi, J.T. Kim, E. Choi, Y.B. Kim, S. Kwon, J. Sun, J. Choo. Retain Vis: visual analytics with interpretable and interactive recurrent neural networks on electronic medical records. *IEEE Trans. Visual. Comput. Graph.*
33. E. Herrett, A.M. Gallagher, K. Bhaskaran, H. Forbes, R. Mathur, T. van Staa, L. SmeethData resource profile: clinical practice research Datalink (CPRD). *Int. J. Epidemiol.*, 44 (3) (2015), pp. 827–836.
34. A. Steventon, S. Deeny, R. Friebel, T. Gardner, R. Thorlby. Briefing: Emergency hospital admissions in England Which may be avoidable and how?, Tech. rep., The Health Foundation (May 2018).
35. C.A. Emdin, P.M. Rothwell, G. SalimiKhorshidi, A. Kiran, N. Conrad, T. Callender, Z. Mehta, S.T. Pendlebury, S.G. Anderson, H. Mohseni, et al. Blood pressure and risk of vascular dementia: evidence from a primary care registry and a cohort study of transient ischemic attack and stroke. *Stroke*, 47 (6) (2016), pp. 1429–1435.
36. C.A. Emdin, S.G. Anderson, T. Callender, N. Conrad, G. SalimiKhorshidi, H. Mohseni, M. Woodward, K. RahimiUsual blood pressure, peripheral arterial disease, and vascular risk: cohort study of 4.2 million adults.
37. Understanding the High Prevalence of Low Prevalence Chronic Disease Combinations: Databases and Methods for Research, https://goo.gl/srsZs2, 4/December/2018.

38. Y. Gal, Z. Ghahramani. Dropout as a Bayesian Approximation: Representing Model Uncertainty in Deep Learning. In: *Proceedings of The 33rd International Conference on Machine Learning, Vol. 48 of Proceedings of Machine Learning Research*, New York, New York, USA, 2016, pp. 1050–1059.
39. A. Vaswani, N. Shazeer, N. Parmar, J. Uszkoreit, L. Jones, A.N. Gomez, L. Kaiser, I. Polosukhin, Attention Is All You Need. In: *NIPS*, 2017.
40. S. Lee. Natural language generation for electronic health records. *Comput. Res. Reposit.* abs/1806.01353.

Index